BUILDING A **BRILLIANT TOMORROW**

BUILDING A BRILLIANT TOMORROW

THE TRANSFORMATION of Inovateus Solar and the Energy Revolution

T. J. KANCZUZEWSKI

Advantage®

Published by Advantage, Charleston, South Carolina.
Member of Advantage Media Group.

ADVANTAGE is a registered trademark, and the Advantage colophon is a trademark of Advantage Media Group, Inc.

Printed in the United States of America.

ISBN: 978-1-59932-730-3
LCCN: 2016959281

Cover design by Katie Biondo.

This book is dedicated to my wife Julia Kanczuzewski, my son Nolan, and my daughter Mara. They give me the love and support I need every day and have been my inspiration in this effort.
I would also like to dedicate this book to my mother, Lyrin Kanczuzewski, who has always been a rock for our family and continues to give me the courage to believe in my dreams.

TABLE OF CONTENTS

FOREWORD

Let's be honest, you probably weren't expecting a racing driver to be writing the foreword for a book titled *Building a Brilliant Tomorrow,* which covers the topic of solar energy. My passion for solar energy and car racing definitely seem to contradict each other at first glance. When I think about why I'm passionate about solar energy, I think that it stems from what I've learned in racing. From the outside, it appears that we climb into the car and go race. What you don't see is all the preparation that goes on behind the scenes. Every practice run we're analyzing massive amounts of data and trying to make the right decisions with car setup to get the best performance in future sessions. The same applies outside of the car. When I think about the effects we're having on the environment today, the technology that we envision in the future, and the associated energy demands, it seems like the easiest setup change decision ever—increase solar!

When I first met TJ and the gang at Inovateus, it was so refreshing to meet a company that shares the same passion for solar energy that I had. It was clear that they not only have the passion necessary to be successful in this industry, they have the engagement, ambition, creativity, and spirit. We came to find out later that these were actually Inovateus core values, P.E.A.C.E (esprit de corps).

Maybe I was a bit naive, but I felt like I already had a good understanding of the solar industry, however, while reading *Building a Brilliant Tomorrow*, I have learned so much more. TJ does a nice job of telling solar's story, whilst at the same time pairing it with Inovateus and his own story that keeps the reader engaged throughout.

Every page of the book contains great advice that can help anyone that aspires to running their own company. TJ provides insight to leading a company on the forefront of the energy revolution, documenting and explaining the ups and downs of the solar industry, as well as going on to predict what the future may hold for the solar industry. I predict Inovateus will play a large role in "Building a Brilliant Tomorrow!"

— Stefan Wilson, IndyCar driver

INTRODUCTION

We are in the midst of an energy revolution. By 2020, the installed solar-energy capacity in the United States is expected to reach seventy-five gigawatts, the equivalent of seventy-five nuclear power plants.[1] Today, there are many opportunities to change how we get electricity in the United States beyond our current electrical infrastructure. Solar is fast taking center stage as a viable, mainstream source of clean energy.

Solar energy is a bridge to energy independence that our country has not seen in its past. It is a key to the future of clean energy, to doing business in a way very different from what we've done in the past. It's the key to building a brilliant tomorrow.

Solar has historically been installed by companies focused on its environmental aspect. Today, it's more often installed because of its economic impact. In fact, on average, the solar-energy capacity in the US is doubling every year.[2] This is just one indication of the dramatic changes the industry has undergone.

Building a Brilliant Tomorrow: The Transformation of Inovateus Solar and the Energy Revolution is about how Inovateus Solar has

---

1 "Solar Industry Expected to Add over 200,000 New Jobs by 2020," Solar Energy Industries Association, accessed January 7, 2016, http://www.seia.org/research-resources/impacts-solar-investment-tax-credit-extension.

2 "Solar Industry Data," Solar Energy Industries Association, accessed August 5, 2016, http://www.seia.org/research-resources/solar-industry-data.

helped businesses and individuals achieve a cleaner future and energy independence. Our story illuminates how a group of people, entrepreneurs, and a business, in particular, have worked together to change the future.

I am a third-generation entrepreneur. My grandfather, Leonard Kanczuzewski, started a logistics company (trucking, warehousing, etc.). Consolidated Services, Inc. (CSI) grew to twelve offices nationwide, and the company's research-and-development team came up with a freight-securement device called Logistick. After CSI underwent a bankruptcy, my father (Tom Kanczuzewski, who was working for his father—my grandpa—at CSI) built a separate company focused on that Logistick invention. My father eventually became the sole owner of the Logistick company, growing it until he passed, after which my sister, Ashley, took over the reins.

As Logistick achieved economic stability, my father, hungry to utilize more of his entrepreneurial spirit, started a business incubator known as Inovateus Development. I was working as a sustainability director for a Fortune 1000 company but left that position to join Inovateus before starting a new company, Inovateus Solar.

Inovateus Solar is based in South Bend, Indiana, a place not particularly thought of as home for a solar-energy company. However, Inovateus Solar has been able to do tremendous things from its Midwest home. Today, Inovateus Solar is one of the country's top providers of solar-energy services, thanks in part to the installation of a project for DTE Energy (Detroit Edison), a Michigan-based utility company. With the ability to offer customers a wide variety of solar-energy products and services, we've built a strong portfolio of customers ranging from Fortune 500 companies to utility companies, universities, municipalities, and the federal government. We do this by using our insights and expertise to educate customers on the benefits of solar electricity. We also make a tremendous impact on customers and others inside and outside our industry—in the US and abroad—by sharing our core values of PEACE: passion, engagement, ambition, creativity, and *esprit de corps*. I'll share more about these values in chapters 4–8.

Building a Brilliant Tomorrow is about the intersection of American business and the energy revolution. Foremost, I wrote this book to share the story of Inovateus; I want others to see what an innovative and growing company Inovateus Solar is. But this book is just as much a how-to story for entrepreneurs and other businesspeople who want to understand how core values can drive a company to achieve bigger and better goals. I hope—from learning the importance of starting any endeavor by asking why, to learning the importance of living core values—other entrepreneurs will get ideas from our story to help them evolve their young companies.

Finally, I want to share my insights on the solar revolution; the next ten years are going to be very interesting in this industry. With advances in technologies and prominent figures getting onboard, solar is fast becoming the energy of choice for many applications.

I'm fortunate to know many industry experts. Two of them have generously shared their insights in this book. My thanks to John Perlin, an internationally known solar historian, and to Rhone Resch, the president of the Solar Energy Industries Association (SEIA), for their contributions.

If I had to use one word to describe the reason for Inovateus Solar's success, it would be *fearlessness*. Inovateus wouldn't be where it is today without the early and ongoing fearless decisions of the founders and the people who've worked for the company.

This book is for anyone who dreams of a better future in this country, for our children and the generations yet to come, and for those pioneers and innovators building a brilliant tomorrow.

CHAPTER 1

EARLY EVOLUTION

The word *revolution* doesn't bring to mind images of peace and tranquility. Usually, it implies some sort of battle is underway and is tied to pain, hardship, challenge, and the changing of the guards. When the idea for Inovateus Development (the predecessor of Inovateus Solar) came about, those were the kinds of words used to describe what the company faced.

The energy industry is undergoing a revolution, and being a part of that revolution hasn't been easy. Running a company on the front lines of battle in the energy industry makes us trailblazers. We see a lot of the changes, almost as they take place. And as trailblazers, we haven't always had a path in front of us for what to do in the face of all this change. But using the tools available, along with the strengths of our team, and by heeding the lessons we've learned along the way, we have enabled Inovateus Solar to maintain its position on the front lines of solar energy. We've not only witnessed some amazing changes in the solar industry; we've actually been instrumental in making some of those changes.

Solar energy is something of a buzzword, having only really permeated the media on a regular basis for the last ten years or so. Before that, solar energy's biggest media blitz was probably back in the late 1970s and early 1980s. Those news reports were largely focused on solar as a solution during the energy crisis and included stories about President Jimmy Carter placing solar panels on the White House and then his successor, President Ronald Reagan, taking them down. Those activities more or less stirred up the political debate that we have in the United States on renewable energy, including support (or lack thereof) by the two prominent political parties.

During the 1990s, many environmental strides were being made from a policy standpoint. Countries such as Germany began implementing renewable-energy standards and creating new financial mechanisms to drive their efforts. In 2004, President George W. Bush implemented a 30 percent investment tax credit for solar energy and similar incentives for other renewable technologies. A few years later, Bush even started a campaign to help get the US off its addiction to oil. When the economy collapsed in 2008, solar and other renewable energies came to the media forefront again as potential business opportunities for the future—options for getting our country out of the recession.

While these events are relatively recent, solar energy has been around for a very long time, going back thousands of years. Today solar energy impacts so many things that we don't think about on a day-to-day basis. Farmers use solar energy to grow and harvest crops, warm-weather resorts use solar energy, and people absorbing the sun's rays are taking in essential vitamins that power their bodies every day. In essence, solar energy powers all of humanity. There are very few species that can survive without it.

While there are various types of solar energy, for the purposes of this book, we're restricting the discussion to solar electricity, specifically photovoltaics.

Photovoltaics are not a new concept. While you don't see them on too many buildings or land throughout the US, organizations and individuals that have sustainability as a mission or energy production and financial investments as a product are starting to install photovoltaics on their buildings or property. The news media broadcast stories about installations of large solar power plants in the deserts of California, Nevada, and Arizona, and about multiple uses for solar in Europe. Of course, solar panels and equipment now being manufactured in China and other Asian countries have put a bit of a damper on the industry in the United States.

Again, that's the world of solar today. But the first photovoltaic installation actually took place in the United States in 1894 in New York City. That solar-energy installation is still running today. Even Albert Einstein had a huge impact on the science of photovoltaics and the efficiency of solar panels. Einstein's studies revealed that light waves emit energized particles, the basis of photovoltaic solar devices.[3] In the 1950s, Bell Labs helped advance photovoltaics and the photovoltaic cell and started making solar panels. Not long after that, companies such as Sharp were starting to do the same thing and advance the product in Japan.

So solar energy is not a new concept, but it's surprising to realize how long it's taken for solar to become a mainstream topic.

My father, Tom Kanczuzewski, was born in 1961 and grew up in South Bend, Indiana. He loved the outdoors, and as a young adult,

3 Glenn Meyers, "Einstein: The Father of Photovoltaics—Part 2," Clean Technica, January 13, 2015, accessed August 5, 2016, http://cleantechnica.com/2015/01/13/einstein-father-photovoltaics-part-2.

he became more involved in business ventures and began to research renewable energy. He helped me do a science project on solar energy while I was in grade school, and when I was a little older, he involved my siblings and me in the local chapter of the Izaak Walton League of America, an organization devoted to protecting and preserving natural resources in the United States. Being involved with that club helped me grow up with a concern for our environment and a passion for environmental conservation. That was just one of many interests I shared with my father.

Around 2003, my father was looking for new business opportunities. His main business at the time was Logistick, a company started in 1991. Logistick was started after my father and his brother developed a freight-securement device while working for the trucking company founded by his father—my grandfather—Leonard Kanczuzewski. Logistick had grown into a very successful venture—the company has about a dozen patents today—and it served as a catalyst for other opportunities.

With Logistick's success, my father started looking into ways he could help the push for renewable energy and maybe even build a business that could focus on that effort. He was always doing cutting-edge research, and he always had the latest-and-greatest technology. Since my father was very interested in computers, I was, as a youngster, constantly on the Internet—even as early as 1992—something none of my friends had access to. As part of the push into renewable energy, my father decided to start a business incubator, first named Inovateus and soon after, Inovateus Development LLC, as a research-and-development effort looking at green construction.

Soon after forming Inovateus Development LLC, my father hired Nathan Vogel and Ed Smith. Nathan had grown up in South Bend and had recently returned home after graduating from the University of Oregon. Ed was a home builder and friend of my father's since high school.

The first concept my father came up with for Inovateus Development was to combine the talents of the three of them to build homes that used various renewable-energy efficiencies. Among the ones that really caught the attention of customers was solar. My father was more interested, at that time, in hydrogen and fuel cells, but solar stuck because it could be adapted easily to the current electrical infrastructure of homes and businesses.

As he undertook research for the company, Nathan came upon an article about George Howard, a Notre Dame professor who had just completed a book about solar pioneer Stan Ovshinsky titled *Stan Ovshinsky and Hydrogen Economy: Creating a Better World.* Among Stan's concepts was that of the clean-energy cycle, upon which he built Energy Conversion Devices. That company was an umbrella for other entities, one of which was Uni-Solar, a solar-energy company that manufactured a thin, lightweight, flexible solar panel that could be installed on homes and businesses.

Nathan reached out to George Howard and arranged for George to meet with my father. The two of them really hit it off in that first meeting. They both loved a lively debate, so while that first meeting seemed a little like a heated argument, it was, in fact, the beginning of a strong business and personal relationship. George was a longtime psychology professor who, at one time, served as president of the American Psychological Association and helped start a world-class psychology department at Notre Dame through the vision of the then-president of Notre Dame, Father Ted Hesburgh. My father, on the other hand, was somebody who went from high school straight into the business world, working for his father at the freight firm

CSI. Although he didn't have a college education, my father knew how to run a business, and he knew how the business world worked.

Although George and my father came from different worlds, they shared the same passion for renewable energy. One of the reasons George wrote a book on Stan Ovshinsky was because he was focused on the psychology behind renewable energy and how it alters the way we think about electricity and consume it. George was concerned about how populations would consume electricity in the future.

After several meetings, George agreed to be a consultant to Inovateus Development LLC in return for help with some of the expenses of publishing his book. George also agreed to help with Inovateus meetings and presentations, from time to time, and he helped forge a product-distribution relationship with Uni-Solar.

By 2006, Inovateus Development had four employees and had begun building homes in southwest Michigan and northern Indiana. Some of the homes were multimillion-dollar properties in Harbor Shores, a new development in Benton Harbor, Michigan. At the time, Harbor Shores was a collaboration between many parties including Whirlpool Corporation, the city of Benton Harbor, and Jack Nicklaus, who helped design the development's world-class golf course.

Tom Kanczuzewski

By this time, I had graduated college and was working at General Growth Properties in Chicago, Illinois.

General Growth Properties is a real estate investment trust that owns, builds, operates, and maintains shopping malls across the United States and abroad. When I worked for the company, it was the second-largest real estate investment trust in the world, a Fortune 1000 corporation of significant size.

I originally went to Chicago in 1999 to attend music school because I wanted to be a professional musician. While I had some success in the industry, music proved to be a lifestyle that didn't lend itself to paying my bills. So I had landed a job at General Growth Properties in 2004. By 2006, I had been promoted a couple of times, and I was beginning to look at ways to share the information on renewable energy that my father was sending to me. I had stayed in very close contact with him since leaving Indiana, in spite of our typical father and son disputes. I was born when my father was nineteen, so we sort of grew up together. He was more than my father; he was a best friend, a teacher, and a mentor.

General Growth Properties was on the cusp of being a Fortune 500 company at the time. I saw that there was enormous potential for the company to do many things, one of which was installing solar panels on top of its malls, which covered more than a million square feet. I saw the company's incredible amount of roof space as a huge opportunity for it to implement solar energy.

I arranged to meet with the CEO of General Growth Properties to talk about my idea. During the meeting, I gave him a copy of George Howard's book on Stan Ovshinsky, and I talked to him about solar energy and the opportunities I saw for the company if it were to install solar. He agreed that the ideas were good, and he wanted to put together a sustainability committee to help with the initiative. I left that meeting ecstatic over the possibilities. I envisioned helping that large company become a force to be reckoned with in terms of solar-energy implementation.

But soon I discovered that implementing a solar system wasn't going to be as easy or fast as I had hoped. Even though the CEO had given the idea a green light, there was a lot of red tape to navigate, including a high amount of criticism and resistance from other leaders in that company.

So all of my excitement began to turn into frustration. I had continued to share copies of George Howard's book with decision makers, along with my ideas about solar, but nothing I did really made the monumental impact I was hoping for at the time.

I grew impatient with the pace of change at such a large company, especially when I really began to feel that the solar industry was about to take off. I wanted to be on the front line of the industry. I was in my midtwenties, and I had the passion and the energy to do big things.

So early in 2007, I pitched to my father an idea to start a solar-energy focus at Inovateus Development, and I offered to lead the charge. I knew that many companies were starting sustainability groups at that time. And while I thought it might be difficult to cold-call a Fortune 500 company and reach top decision makers, such as the president or the chief financial officer, I thought we could succeed in reaching people at the level of vice president of corporate responsibility or vice president of sustainability. Those were fairly new positions at the time in corporate America, and the people in those roles were open to hearing about products and solutions to achieve energy efficiency, cost savings, and a positive environmental impact.

My father was not so keen to move forward with the idea despite being happy to see I was passionate about solar and proud that I had ventured off on my own. He was glad I had landed a job at General Growth Properties and had gone from being a music student to working at a big company where I had received several promotions in a relatively short amount of time. Again, I was disappointed, but I refused to give the idea up.

A few months later, I met my father and the rest of my family in St. Augustine, Florida, for a vacation. On that trip, I woke up one morning and decided to again pitch my father on the solar idea as we sat by the pool, having some coffee.

Even though Inovateus Development was making some headway in research and development, my father tried once more to turn me away from the idea of joining the company and leading a solar initiative. That's when I finally said, "No, Dad. I've thought long and hard about this. I don't feel like I'm using all my skills and talents in my current job. I want to work in solar energy. It's something I believe in. It's something I'm passionate about. If you don't move forward on this idea, I'm going to go work for a solar-energy company."

That perked him up a little, and he asked me what I meant by working for a solar-energy company. I told him about solar-panel manufacturers and about the small number of solar firms I had come across while working at General Growth Properties that were looking to do installations for Fortune 500 companies and utilities.

When he saw that I was very committed to my beliefs, he decided to consider my idea. However, he told me I had to put together a written plan and come to Indiana to present it to him. So I did just that.

I put together my idea and had a meeting with him at the beginning of June 2007. After the presentation, he didn't commit to me right away. Instead, he thought about the idea for a couple of days, but ultimately he asked me to join Inovateus Development as a vice president of sales in July 2007.

Before leaving my job at General Growth Properties, I took a day off and went to South Bend to meet with my father and George.

They wanted me to ride with them to Rochester Hills, Michigan, to meet with Stan Ovshinsky. At the time, my father and George were thinking about investing in a fuel-cell business, and Stan had technology and patents and even a fuel-cell division of Energy Conversion Devices that could be included in the venture.

So, my father's solar interests, which George Howard had sparked, were also driven by the efforts of Stan Ovshinsky. When Stan started Energy Conversion Devices back in the 1960s, his mission was basically to create a clean-energy loop to generate electricity from solar and store it. Electricity could then be used in real time and also stored in fuel cells. The idea was to continually generate and store electricity through a loop system.

Today, in terms of storage, some of the most prevalent technologies use a lithium-ion type of battery. However, Stan was more focused on fuel cells and hydrogen storage, which interested my father. One of the challenges was that the existent solar-energy infrastructure needed major overhauls to accommodate fuel cells and hydrogen storage. Those are two pieces of Stan's mission that still have a long way to go. Lithium-ion batteries and other technologies have been the most cost competitive and efficient to this point.

On that trip to Michigan to see Stan, I rode with my father and George and we discussed solar and renewable energy throughout the entire trip. When we met with Stan, who was in his mideighties at the time, he shared with me his story and how his company had got to where it was at that time.

In the meeting, my father and George told Stan about the new company they wanted to start, and they invited Stan to be involved. To convey his interest, Stan pulled out his checkbook, wrote a six-figure check, and handed it to my father. Ultimately, they didn't move forward with that initiative (my father sent the check back to Stan).

But my father's solar ideas had an influence on Stan's mission, started over fifty years earlier, to create a business to implement the marriage of clean-energy technologies.

There is one thing I remember very clearly about that meeting, and to this day, I carry a picture of the moment in my mind: I looked at the faces of the people in that room—my father, George, and Stan—and I thought, *This is what I'm supposed to be doing*. In that instant, I felt whole. I felt I was where I was supposed to be. I truly felt solar was my mission.

Left to right: Tom Kanczuzewski, George Howard, Stan and Iris Ovshinsky, Nathan Vogel, and John Cernak

After that meeting, I left my corporate job and went to work at Inovateus Development, which had just become a distributor for the flexible, lightweight product produced by Uni-Solar. Back then, Inovateus Development didn't have a finite business plan but rather a hodgepodge of different ideas. One of those ideas concerned building residential homes and businesses in which the Uni-Solar product would be installed. In my first year with Inovateus Development, I worked with the team to sell real estate and Uni-Solar products, and on the side, I wrote a formal business plan to take Inovateus Development into the solar arena.

I finished the plan in September 2008. Although we had approached other investors with the plan, my father insisted on retaining a 51 percent ownership in the company, with George Howard owning 44.5 percent and my grandfather, Leonard, owning 4.5 percent.

We started Inovateus Solar LLC on October 1, 2008. We had a business plan to be a solar-energy product distributor, and we set off to accomplish what we had put down in writing.

Two weeks later, the economy collapsed, and across the US, some of the other renewable companies that had begun to take off only a few years prior got hit very hard. The situation changed the renewable-energy landscape. What a great time to start a new company!

Inovateus Development continued to do residential construction, but Inovateus Solar focused on meeting with companies at the Fortune 500 level. We worked to educate their team members on solar energy while trying to sell them our solar-energy products. With the economy on shaky ground, some of those companies were struggling to survive, including my old employer, General Growth Properties. The company I left behind—to the sound of snickering coworkers who laughed at me for leaving to work with my father in the field of solar energy—nearly went under in 2009.

Then, early in 2009, Inovateus Solar got its foot in the door with many large companies. One of those was Procter & Gamble, a company we had already pitched as representatives of Inovateus Development. We also diversified our product offering beyond Uni-Solar to include other solar-panel manufacturers and electrical-component providers.

One reason for the diversification of the product was the Uni-Solar team's lack of support. The company had started to focus on working with roofing manufacturers such as Carlisle. Stan Ovshinsky had retired from the company and the shareholders and board of directors brought in a new regime focused on driving profits. The culture of the company changed dramatically, and loyal Uni-Solar fans and customers such as Inovateus were starting to get the cold shoulder.

At Inovateus Solar, the conversations still centered on selling only the products; installation was not a part of the company's offerings.

However, as we started to make some of our first sales, clients began asking us for designs, project management, installation, and other services. Companies were happy to buy our products, but they had no one to install them. Qualified, quality installation services were few and far between back then, and companies struggled to find them.

Recognizing a need, we adapted the Inovateus Solar business plan to what clients were asking for; we became not only a product provider but also a project-implementation service.

We also had to develop the team at Inovateus, so we started hiring engineers and designers. We expanded our sales effort, and we even brought construction project managers onboard.

Soon, we began to stumble into some sizeable orders. By the end of 2009, we were getting good-sized, rooftop, solar projects in New Jersey, and even large product orders around the world in places such as Qatar, Greenland, Australia, Fiji, and China. By using our strengths, building relationships, and sharing our passion for solar energy, we began to build a name for ourselves in the solar-energy industry. I think many manufacturers and other companies really were refreshed by our strong belief in the importance of solar energy in the United States and the world as a whole.

By 2010 Inovateus Solar had reached $25 million in sales. We went from zero to $25 million in sales in a year and a half.

CASE STUDY

In 2010, Inovateus Solar contracted with Masser Farms Realty and began construction of a 1,040-kilowatt, ground-mounted, solar system at the company's seventh-generation potato farm in Sacramento, Pennsylvania.

We worked on the project with Keith Masser, which gave us a chance to learn about that amazing family business. Solar was something the company was familiar with since the company's products—potatoes—rely on solar energy to grow.

The system consists of Schott Poly solar panels, in combination with Schletter racking and two Schneider Electric Xantrex GT500 photovoltaic inverters. The system is one of the largest, privately owned, solar installations in Pennsylvania and generates over 1.6 million kilowatt hours of pollution-free electricity.

The system will eliminate 2,396,159 pounds of carbon dioxide and 2,115 pounds of mononitrogen oxide and produce enough energy to power 150 homes every year. This solar system is the beginning of Masser Farms' alternative-energy initiatives.

The Masser Farms installation is a great example of how a business with a long legacy can adapt within an energy revolution—and they've been around for a few—and set an example for others.

Masser Farms Installation

CHAPTER 2
THE SOLAR STORY

The solar industry and our capacity have, on average, nearly doubled every year since 2008. With every project Inovateus Solar does, we see more and more the breakdown of the old utility structure as businesses and individuals yearn to become energy independent, financially healthier, and more focused on environmental responsibility.

Yet the solar industry is still trying to justify its value. It's interesting because I see it in some ways paralleling the IT revolution: no one believed in some IT companies that have since changed the way the world operates.

Back around 2004 and 2005, when Inovateus Development was in its infancy and was researching the various renewable-energy technologies, the future of electricity was beginning to gel, and solar was definitely starting to become front and center. Ultimately, we decided to pursue solar because we knew solar had the greatest potential to lead us into a future with cleaner energy since it can coexist with the current electrical infrastructure and does not require the drastic overhauls needed by other potential sources.

We saw that things were changing. Even back in 2005, when we were first looking at solar, the industry was doubling every year. We saw there was growth in the industry, and when Inovateus Solar started in 2008, that growth continued. It was clear to us that solar energy was at the forefront of breaking down the old utility structure: when businesses installed solar energy at their facilities or individuals installed it at their homes, they could connect to the grid, but they could also start to become independent. That independence is what really added fuel to the solar revolution.

When we started Inovateus Solar, we were definitely pioneers in the Midwestern market. I knew it then, but it really hit home in 2015 when my brother Tyler and I went to the inaugural Solar Pioneer Party in Humboldt County in northern California. I'll talk about this more in an upcoming chapter, but at that event we met individuals who had been working in the solar-energy industry since the 1970s and earlier. That party really opened my eyes. It made me pause to learn a few things about the history of solar energy.

The industry started to see some real breakthroughs in the late 1970s and early 1980s. However, while a lot of change was happening in California and along the West Coast, an area of the country that tends to be very progressive, the market there was already developed and folks there were pretty well educated on solar.

It took some time for solar to evolve in the Midwest. In 2008, when we were going to market as Inovateus Solar LLC, a lot of the companies and individuals that we met didn't have any real knowledge or a very good understanding of solar energy, so we spent a lot of time trying to justify solar's value by focusing on education. Since we were based in South Bend, Indiana, we focused largely on a five-hundred-mile radius around the city. We spent those early years trying to educate people about solar energy, and we tried to bring them up to speed on solar energy's industrial status. We realized it was our duty to help inform the general public about the environmental and economic benefits of solar.

In addition to establishing a network of customers and getting our foot in the door with the companies we'd done business with, we had to reach out to industry experts. That meant joining many different organizations, one of which was the Solar Energy Industries Association (SEIA).

We were also foundational in starting what's called the Indiana Renewable Energy Association (INREA). At that time, there was no network within the state of Indiana to help educate the general public. There were other organizations—such as the Hoosier Environmental Council, in which we have been active since our beginning—but they focused on environmental issues or other, broad types of renewable energy.

Through the Indiana Renewable Energy Association, we began to be instrumental in educating the Midwest. Since we're based in Indiana, we obviously wanted to start our efforts to educate the public in our home state. We became involved with the Midwest Renewable Energy Association (MREA), which is based in Wisconsin. We started attending a lot of the events the organization sponsored, including a popular renewable-energy fair—complete with workshops, demonstrations, speakers, exhibitors, and so on—during the summer months.

In 2008, solar energy, from a mainstream perspective, was still in its grassroots stages, despite being something the human race has been trying to tackle for over six thousand years.

Obviously, in my role at Inovateus, I see things from a certain perspective. But there are a number of experts in the industry who are also telling the same story.

For instance, international solar expert John Perlin published a book in 2013 that discusses in-depth this centuries-old energy source. The book, titled *Let It Shine: The 6,000-Year Story of Solar Energy*, is a must-read for anyone who wants to understand the history of solar energy.

I had the opportunity to meet John when I was at the Solar Pioneer Party in California. While I was very excited to be in the company of individuals who really helped kick-start the US solar-energy industry, a short conversation with John helped me realize that what we've lived through is truly a small snapshot of only the most recent history of solar energy in the United States.

In fact, solar-energy technology is something that people have been working on in the United States for more than one hundred years. Experts and scientists such as Albert Einstein were major contributors to the advances that we enjoy today. Most people don't realize that Einstein won the 1921 Nobel Prize for his experiments with photovoltaics, specifically his discovery of the law of the photoelectric effect. Companies such as Bell Laboratories in the US and Sharp in Japan were busy advancing solar cells in the 1940s and 1950s. In 2012, me, my brother and coworker, Tyler Kanczuzewski, and coworker Nathan Vogel toured the Kyocera corporate headquarters in Kyoto, Japan, and saw the work they had started in 1975 during the energy crisis, which involved the development of solar cells for the general public. While 1975 seemed long ago for the three of us, it is like only yesterday in the history of humans and solar energy.

Looking further back, John's book touches on the way that humans were already trying to capture solar energy in developing villages six thousand years ago. I found it a very interesting read, and with John's consent, here are some takeaways from *Let It Shine: The 6,000-Year Story of Solar Energy*.

LET IT SHINE: THE 6,000-YEAR STORY OF SOLAR ENERGY TAKEAWAYS:

- As far back as six thousand years ago, people understood the importance of using solar energy to heat the home and building their structures to take advantage of the seasons and the orientation of the sun.
- Facing a home to the south is the first principle of using solar in construction, and to date, excavations have shown that China has used it longer than any other civilization.
- Sometime around 2,000 BCE (four thousand years ago), the ancient Chinese developed a device called the "gnomon," which was a stick or other item that was planted in the ground perpendicularly to record with shadow the movements of the sun. This device was first used to situate buildings by the Zhou dynasty, which began approximately 1046 BCE.
- The second-century poet, Ban Gu, witnessed the power of the sun at work in early Chinese solar architecture. He wrote that, when the palace's south-facing door was open in wintertime, "the sun's radiance would flare brilliantly into the palace, heating the rooms inside." This brightly lit entryway was known as the Door of Established Brightness.
- To combat the intensity of the sun during the hotter months, structures were outfitted with overhanging roofs; excavation of architecture in northern China from civilizations more than six thousand years old has revealed some of these architectural elements.

- Doors were also used to keep rooms cooler by blocking the summer sun's rays.
- Entire cities, in fact, were laid out on a grid to take advantage of southern orientation.
- And over time, southern orientation came to symbolize brightness and even higher knowledge; elders and respected family members were seated in the southwest corner of a room, and emperors faced south as a way of shunning the darkness and embracing enlightenment.
- By the time of Socrates (470–399 BCE), building with solar in mind was commonplace.
- When some 2,500 people broke from Athens in 432 BCE and moved to Olynthus, a new district known as North Hill was established in the same manner as the Chinese, with perpendicular streets that allowed for all houses to face south.
- No matter the terrain, ancient architects took advantage of solar, as evidenced by the Greek city of Priene, built on the difficult and steep slopes of Mount Mycale and Delos, an Aegean trading center built on rocky-island terrain. The Roman architect Vitruvius, (90–20 BCE) was a military engineer in Greece, and would write about what he found there in his books on architecture, adding his own recommendation that winter dining rooms should face the setting sun in winter, and summer dining rooms should be oriented to face the north.
- Romans also began using glass or glass-like materials, such as mica, to cover openings in order to trap solar heat

in a room, which ultimately led to early greenhouses. One notable structure that took advantage of window coverings to create a "solar furnace" was at a seacoast retreat of the wealthy Roman statesman Pliny the Younger (61–113 AD). Romans even managed to architect their bathhouses to take advantage of solar to the point that bathers often "broiled" by late afternoon.

- Over time, solar energy became useful for other purposes including heated floors and "burning mirrors," the latter of which were used to "concentrate the sun's rays onto an object with enough intensity to make it burst into flames in seconds"—a useful tool in the days before matches and lighters. The earliest of these tools dates back to the Chinese Bronze Age, but the idea was reignited later because burning mirrors were used by the Greeks to light the flame signifying the start of the Olympic Games. The burning-mirror device can be credited for the belief that the sun equated to fire to both the ancient Chinese and Greeks.

- Using solar rays to create powerful weapons of war was definitely considered over time, but despite all efforts, no such weapon was ever successfully produced. The effort to create larger, more impactful tools, however, did result in a ten-foot diameter, wood-and-brass, concave mirror by the German craftsman Peter Hoesen. This tool could cause objects to burst into flame and/or melt in seconds. Gunpowder made the mirror-as-weapon-of-war idea obsolete, but not before Isaac Newton created a device made of seven mirrors that "vitrifies brick or tile in a moment," according to a witness.

- In the seventeenth century, the use of solar energy for horticulture began to take hold through greenhouses and "fruit walls," the latter of which took advantage of the sun's rays on a wall made of brick or other material.
- Tweaking this tool to make better use of the sun eventually led to sloping of the walls and then glass coverings, which led to the "age of the greenhouse" in the eighteenth century and to the construction of attached conservatories by the end of that era.
- Another invention during the eighteenth century was the solar hot box, a minigreenhouse of sorts that trapped the sun's rays inside. European naturalist Horace de Saussure's experiments using his hot box in the Alps and on the Plains found that the "sun shines with almost equal force at higher and lower elevations"—temperatures inside the box rose to nearly the triple digits regardless of altitude and regardless of the temperature outside the box.
- With the advent of the Industrial Revolution, solar began to be explored as an energy source to power motors, although an early device powered by the sun was created in the first century BCE by Hero of Alexandria. The device consisted of two stacked containers connected by a tube and siphon system. The upper container was a sphere partially filled with water. When the air inside the upper container was heated by the sun, it expanded, thereby forcing the water out through the tube and into the bottom container.
- The solar-siphon idea was little more than a toy used to create sound devices until Augustin Mouchot, a

mathematics professor at the Lycee de Tours in France, began experimenting in earnest to create solar devices for practical purposes. He was concerned about the finite source of fossil fuels in use at the time, primarily coal. Using the glass heat trap and burning-mirror concepts, Mouchot created solar-reflector technologies that resulted in several successful solar-powered inventions: a still, an oven, and a pump. Mouchot obtained his first solar patent in March 1861, and five years later, he created the first solar-powered steam engine. In the ensuing years, Mouchot succeeded in creating workable boilers and other devices, but by 1880, other forms of generating electricity were proving to be more efficient, and Mouchot left his pursuit of solar research in the hands of his assistant.

- Solar-energy research emerged in America in the mid-1800s with inventor and engineer John Ericsson, who was also concerned about coal consumption. Solar evaporation used in the production of salt was the most fuel-saving application of the time and was a process used around the world. Ericsson went to work on a solar engine, ultimately producing a solar hot-air engine that in the end wasn't commercially viable. He made headway in reducing the costs of solar reflectors and was on the verge of selling solar pumps to California farmers when he passed away, taking many of his solar secrets with him. Following in Ericsson's footsteps, Aubrey Eneas, and English inventor and engineer, over time built a solar motor that went on display on an ostrich farm in southern California in 1901. The motor powered a pump that drew an impressive 1,400 gallons of water per minute from a nearby reservoir to irrigate a three hundred-acre citrus grove.

- The efforts of the early pioneers all had several things in common: the collection of solar meant the use of ultrahigh-temperature collectors, which led to large, inefficient heat losses. The complexities of the equipment made them expensive and vulnerable to Mother Nature's worst—they had to follow the sun to collect energy, and they didn't work on cloudy days.
- To combat the downsides of earlier inventions, in the late nineteenth century, inventors like Charles Tellier, a French engineer known as "the father of refrigeration," began to experiment with low-temperature solar collectors. These devices relied on low-boiling-point liquids like ammonia hydrate and sulfur dioxide.
- Following Tellier were two American engineers, H. E. Willsie and John Boyle, who created a solar power plant built on a simple hot-box collector that heated water enough to power a low-temperature engine. Their invention turned into the first twenty-four-hour solar-powered generator with the aid of an auxiliary boiler that was powered by conventional fuel that took over on sunless days and overnight. Willsie and Boyle opened a plant in Needles, California, in the Mohave Desert, in the early 1900s, which after two years proved to weather well. Repeated improvements ultimately led to the first—and long coveted—solar-energy storage system, which allowed the plant to operate overnight from solar energy that had been collected during the day. The solar plant

was economically superior to coal, the other fuel source available in the southwestern part of the US at the time, until a few years later when gas power was introduced to the area.

- The late 1800s through the early 1900s saw the introduction of the solar water heater. In 1909, the appliance was revolutionized by William J. Bailey, who created a unit that supplied solar-heated water around the clock, but this too was ultimately phased out for the sake of gas water heaters. In the 1920s, solar water heaters made a comeback, and although the Great Depression led to a slump in the market, in 1935, under the New Deal, a wave of new construction in Miami produced new opportunities for roof-mounted solar water heaters. By the early 1960s, costs of materials in the wake of World War II, combined with tenth-year failures of installed systems and increased demand for water usage in a newly affluent society, led to a decline in the Florida solar industry, although solar water heaters continued to be used in other parts of the world.
- The refinement of double-pane insulating glass by the Libbey-Owens-Ford Glass Company led to the construction of a solar-housing neighborhood in Chicago in the 1940s for developer Howard Sloan. The homes were designed by architect George Fred Keck, who eventually convinced the Illinois Institute of Technology to conduct a year long test of solar on a home known as the Duncan house.

- Keck solar homes with concrete floors held the heat longer, acting as early solar-storage systems (although solar-storage systems date back to the 1880s, when Edward Sylvester Morse, a Massachusetts botanist at the Essex Institute, developed a system for storing heated air).
- Heat storage was also the goal behind the black walls incorporated into designs by Arizona architect Arthur Brown.
- In 1938, research began in earnest on solar collectors, when engineers at the Massachusetts Institute of Technology (MIT) began a two-decade look into collectors for heating homes.
- The MIT team built four homes with varying degrees of success, ending the experiments in 1962.
- The photovoltaic effect—the flow of voltage or an electric current through a solid material exposed to light—was first explored in the late 1870s using selenium bars. But since the material didn't generate energy by relying on heat—which most other solar devices did at the time—it was dismissed early on as a viable material for further research.
- It was another seventy years before the first practical solar cell appeared, created by accident by two Bell Laboratories' scientists who were working on silicon transistors.
- The technology was too expensive for everyday use, so it was first used in 1958, when the solar battery found

its way into space as a power source for the Vanguard I satellite.

- Costs had kept the solar battery aloft until 1968, when industrial chemist Elliot Berman produced a film-like cell that caught the eye of Exxon executives. Although still too expensive for home use, the photovoltaic modules brought considerable cost savings when replacing nonrechargeable batteries on navigational warning systems on oil platforms in the Gulf of Mexico.
- Soon photovoltaic cells were being used for a variety of applications around the world. The cells proved particularly useful for remote applications, such as powering microwave repeaters for the telecommunications industry.
- The glut of energy sources—cheaper oil, gas, and electricity—for the three decades following World War II ultimately led to a decline in solar interest.
- The push for atomic power further added to solar taking a back seat in the field of energy. But into the early 1970s, solar enthusiasts continued to tinker with solar ideas, leading to some successes such as a solar pool heater and the development of what would become known as "passive solar."
- Passive solar is a system void of pumps or fans that relies instead on the concept of making use of solar energy by allowing the sun rays in and then extending the use of the energy by trapping the heat inside.

- The system uses key building materials such as insulation and concrete floors, along with strategically constructed architectural elements such as south-facing glass, window coverings, and doors.
- Solar reentered the limelight briefly in the late 1970s thanks to President Jimmy Carter, who installed solar on the White House during his administration because of the oil embargo of the early 1980s. But within a decade, the price of fossil fuels again bottomed out. Compounded by the end of a subsidy for solar water heaters in the mid-1980s (which had made a resurgence in the US in the 1970s) and the dim view of solar by the Ronald Reagan administration, the industry in the US lost its momentum. Although not commonplace in the US, solar water heaters have continued to be used with varying levels of success around the world; some fifty-four million had been installed worldwide by 2009.
- Rooftop photovoltaic installations have also been better supported by government initiatives in other countries around the world; by 2002, Japan was a world leader in grid-connected systems and in the manufacture of solar cells, and by 2013, more one billion watts had been installed.
- In 2012, nearly half of Germany's electricity was photovoltaic generated.
- Today, advances in photovoltaic materials are continuing, with a myriad of choices now available in the market. These varying materials have made solar a more cost

effective, more efficient solution for a range of uses around the world.

- Energy storage is also advancing quickly, and it's feasible to envision a day when large areas will no longer be solely reliant on the grid.

Obviously, the history of solar energy is a very long one. It's really one that began the first day the sun shone on Earth. The sun has had the biggest impact on our planet, and when you get right down to it, it's had the biggest impact on all sources of energy that we've been consuming over time. Over the last 150 years, we've watched the mainstream transformation of how we generate electricity. While we still generate electricity using hydropower created by dams and turbines, a major part of the electricity we generate comes from coal, a fuel derived from fossils that originally were plants that got their power from the sun. Ironically, coal has now proven to be one of the greatest contributors to global warming and environmental issues.

I really saw the effects of coal firsthand when I went with Nathan Vogel and my brother Tyler to visit solar manufacturers in China and Japan back in 2012. In Beijing and Shanghai, I was able to actually see the air pollution generated by the coal-fired plants in those manufacturing-intensive areas. It was mind-boggling to see the effects of air pollution, to see people walking around with masks on because it was dangerous to breathe the air. In the six days we were in China, I never saw the sun without a heavy haze clouding it. Later, out of curiosity, I looked online for satellite images, and it turns out you can actually see the smog from space!

Air pollution over China

But coal is obviously a major form of energy. Because our company is based in Indiana, one of the top coal-producing states in the country, solar energy was a challenge to share with many people in and around our local area in the early years of Inovateus Solar. Southern Indiana, in particular, has a lot of coal mines, and there are some very large coal companies in our state. So, operating in a state with a coal economy came with additional challenges beyond those a start-up might face: we had to educate people to power their businesses in a different way, as well as to understand the economic impact.

Around the time Inovateus Solar started as a company, there was a major push in corporate America to look at renewable energy, sustainability in particular, and companies were starting to see value in unused space on the top of their buildings. I had seen this at my previous employer, General Growth Properties, when I worked in that organization's sustainability group. Renewable energy was a low-hanging fruit, a way in which companies could become energy independent. To create clean electricity, organizations could maximize roof space, which was then merely a chunk of real estate used for nothing other than to keep the building dry and to house ventilation systems.

The United States had seen some of the success that European countries were having with solar-energy programs. For instance, Germany initiated a feed-in tariff program for solar energy back in the 1990s. A feed-in tariff essentially meant you could build a solar-energy system that could be connected to the grid, and the utility would pay per-kilowatt-hour for the electricity produced. Instead of having to build dozens of power plants, the utility companies could take advantage of all the little power plants their customers built and buy electricity from these plants all over the country.

After Germany implemented the program for its commercial customers, the country welcomed a manufacturing industry. Since the country was embracing solar, the manufacturers that sprang up also began implementing solar in their operations. Seeing the success of the program in Germany, other countries such as Italy and Spain followed suit. Later, countries such as Japan became major players. As a manufacturing center, Japan had developed and advanced photovoltaic technologies in the 1950s and beyond.

Meanwhile, in the United States, President George W. Bush made a statement in 2006 that America was "addicted to oil." He announced that the addiction needed to end, and he had a multistep process for making that happen. This is all interesting to note, considering that the Bush family was heavily involved in the oil business (it had roots in Texas and had negotiated many deals with Middle Eastern oil companies). But around that time, President Bush also put together with Congress a tax credit for businesses and individuals who were looking to invest in solar-energy technologies. After he did that, the industry started to get a little more media recognition in the United States.

In 2008, when the economy tanked, the growing renewable-energy and solar-energy-specific industry began to get some positive news that media outlets could grab onto. The media picked up on the story of solar energy and the fact that the industry was doubling in growth every year. News stories talked about the great opportunities for the United States to get more involved in the manufacturing of solar-energy technologies to support the growing industry.

One of the companies seeing phenomenal success at the time was Stan Ovshinsky's Uni-Solar, which had manufacturing facilities in the Detroit and Greenville, Michigan, areas and was expanding internationally. There were also a number of US solar manufacturers opening and ramping up.

With the economy being what it was in late 2008, a number of manufacturers in many industries began to struggle. Indiana is a manufacturing-intensive state, and one of the industries hit hard was the manufacture of recreational vehicles (few people had the discretionary income to buy an RV at the onset of, or during, the Great Recession). To stay afloat, the state's manufacturers had to look at embracing a new line of manufacturing and, to us, that meant solar energy. So Inovateus and Uni-Solar were actually involved in a number of conversations to bring solar manufacturers to Indiana.

But as the solar industry started to grow into the global marketplace, Asian manufacturers, particularly in China, started to see the opportunities that were available in solar energy. In 2010 and 2011, innumerable manufacturing plants were brought online in China, creating overwhelming competition for solar products on American soil. US makers were undercut by as much as 50 percent. Chinese manufacturers started to win a lot of the business in the US because they could sell cheaper solar panels and products than US manufacturers could. As a result, a lot of the US makers that had just gotten started and that had had a lot of great publicity just a few years earlier started having financial difficulties.

One of those companies was Solyndra, which was headquartered in California. The company produced a solar panel wrapped in a tube that was able to generate electricity in a 360-degree area. It could go on top of a white roof and collect energy from the sun and also from the light that bounced off the white roof.

Solyndra was able to get a US Department of Energy loan guarantee, something that was ushered in by George W. Bush at the end of 2008, as he was leaving the office. When President Barack Obama came into office in January 2009, the Solyndra proposal and loan guarantee hadn't been officially approved, but he eventually approved it.

Solyndra's product involved an intensive manufacturing process, and just as it was getting started, the market changed, which ultimately led to the company's demise. The company could not bring down its costs of manufacturing fast enough to compete with the market prices driven by the Chinese manufacturers. The company went bankrupt, and it became a very hot topic during political debates, especially during the 2012 presidential race between the Democratic incumbent, Obama, and Republican challenger Mitt Romney.

So even with its ups and downs, solar is a fast-growing US industry. But as you have no doubt discovered from what you've read so far, it has a very feast-or-famine existence that I sometimes like to call the solar-coaster. Along with its challenges to gain a foothold, it's an industry that's gotten a healthy mix of good and bad publicity.

Today, in 2016, most of the publicity is very positive. In fact, there have been some very positive, major changes. I was fortunate to have had a conversation about the industry with Rhone Resch just before he left his role as president of the Solar Energy Industries

Association (SEIA). I asked him where solar is today and what some of the opportunities for the industry might be in the future. Here's my conversation with Rhone:

Me: What's the US solar-industry growth per year?

Rhone: The US solar industry has grown from installing just one hundred megawatts (MW) per year in 2006 to more than 7,000 MW in 2015. By 2021 the US solar industry will deploy more than 20,000 MW, bringing total capacity to nearly 100,000 MW, thanks to legislation passed by Congress that extends the investment tax credit. This means, by 2021, there will be enough solar energy to power twenty million American homes, which translates into well-paying jobs for hundreds of thousands of Americans. By the end of 2020, the solar industry will employ roughly 420,000 workers, more than double the number of solar workers today.

Me: What is solar's place in the energy revolution?

Rhone: As the grid adapts, solar energy is uniquely positioned to succeed. Solar is now taking a bigger piece of the generation pie, especially for peaking generation. It's both cost effective and reliable, and the world is noticing. Solar energy now has firmly secured its place at the table with the nation's other major electricity producers, and if we are to have a realistic hope of continuing greenhouse gas reduction, greater deployment of twenty-first-century solar technologies—both in the United States and around the world—is absolutely critical.

Me: How has the pricing structure of solar industry changed in the past ten years?

Rhone: Over the last decade, the cost to go solar has dropped dramatically. The average price of a solar installation has fallen by more than 70 percent. Utility-scale solar projects now regularly secure contracts at price points of $0.04/kWh or less, similar to fossil-fuel prices. With new financing options, and more capital moving into the industry, it has never been cheaper or easier to install solar on your home or business. However, while costs have sharply declined, they can be pushed down still further. The extension of the investment tax credit (ITC) will help make that happen.

Me: How has the solar industry grown in the Midwest?

Rhone: Just like the rest of the nation, solar in the Midwest has steadily grown. There is now nearly 400 MW of solar in the Midwest, almost double what there was just two years ago. Looking ahead, we expect that growth to continue. Between 2016 and 2020, nearly 2,000 MW of new solar capacity is expected to be added in the Midwest, enough to power roughly 250,000 homes.

Me: What are some significant stigmas associated with solar?

Rhone: There is no question that we have some challenges, and some of the "stigmas" are ginned up by opponents of the solar industry. In 2016 we are implementing a consumer-protection process—www.seia.org/policy/consumer-protection—that addresses some concerns that have been raised in communities. Technology advances and innovations also are going to serve our industry well. As storage capabilities and battery technology advances, solar will increasingly be able to provide electricity anytime, night or day—something it already can do through concentrating solar power paired with storage. Any "stigmas" that may exist won't stick as solar becomes less expensive and more pervasive as the energy source of choice for homeowners, communities, and commercial users.

Me: Where is solar headed in the future?

Rhone: Solar is the planet's most abundant source of energy and offers all Americans clean electricity that can be built at scale. Solar works in all fifty states, has zero carbon emissions, creates more jobs per megawatt than any other technology, and can be deployed cost effectively and quickly, all while improving grid reliability. When you combine all of those benefits with the recent ITC extension, the future of solar energy is looking brighter than ever.

Me: And what role does energy storage play in solar's future?

Rhone: There's no question, innovation in storage will help to drive solar deployment in the US to new heights and open entirely new revenue streams. Solar coupled with storage will be a game changer, helping address questions of intermittency, load shifting, and demand response. Gone will be the doubters and their questions about what happens when the sun doesn't shine. For American homeowners and businesses, solar will become an around-the-clock source of dependable energy.

When Inovateus started in 2008, Indiana was in the bottom five of the fifty states in terms of solar installations. The largest installation at that time was a twenty-kilowatt solar installation. Just seven years later, the largest installation in the state is twenty megawatts at the Indianapolis International Airport, and there are many other large installations, such as those atop Duke Realty buildings, which I'll talk about in an upcoming chapter.

So we've been able to see firsthand the monumental changes that have happened in Indiana, a state that was far more supportive of coal than of solar. Now things have changed so much just for the utilities in Indiana and the Midwest that some of the old coal plants are being decommissioned, although in part because it's too expensive to retrofit the plants to current Environmental Protection Agency (EPA) standards. Plus, a lot of the plants are just reaching the end of their service life. The same goes for many nuclear power plants, including Cook Nuclear on Lake Michigan, nearly thirty miles from our headquarters in South Bend, Indiana.

Utilities are being pushed to adapt their business plan to meet environmental and health regulations. At the same time, the cost for solar energy has dropped incredibly over the last ten years, reaching the point where even without an EPA-driven initiative, utility companies would probably bring in solar anyway just because it's cost competitive with conventional energy or electricity generation.

When we started, most of our projects were on the East Coast in states such as New Jersey because there was a high cost for electricity there. New Jersey also had renewable-energy credits (RECs), which encouraged solar. In Indiana, the cost for electricity has gone up, maybe 20 percent or more, while the price for solar energy has dropped 80 percent. So while most of our projects were outside Indiana in the past, now we're going to focus on areas closer to home.

The solar revolution is real. We see it firsthand because it's what we do; it's what we live and breathe at Inovateus Solar. The opportunities moving forward in the future are just staggering. Even though we've seen monumental changes, we're still only scratching the surface here. We're just now working on some of the first solar projects that utility companies have chosen to implement, and we're starting to work with businesses and individuals who want to integrate solar energy into their everyday lives.

Back when we started, there were probably about five hundred companies within the United States that were involved somewhere in the solar-energy value stream, whether they were a manufacturer of solar panels or small components that went into solar panels or other pieces of the system, such as inverters, racking systems, monitoring, and so on. I'm also including banks and government entities such as the National Renewable Energy Laboratory (NREL) in that count.

Today I would estimate that there are probably fifty thousand or more companies in the United States that, somehow, are involved in the value stream in solar energy. As more solar energy is deployed, that number will continue to rise. Today's solar value stream includes insurance companies, surety providers, investment firms, pension funds, hedge funds, private-equity firms, accessory companies, battery companies, roofing manufacturers, ESCOs, utilities, cooperatives, municipalities—and the list goes on.

We're also seeing the technologies improving, becoming more integrated. For example, we're seeing inverters becoming more efficient. We used to install inverters rated at six hundred volts, and in an upcoming chapter, I'll talk about how we were one of the first to use one-thousand-volt inverters. Today we're moving to 1,500-volt inverters and panels.

Financing mechanisms are also improving, and prominent figures such as billionaire entrepreneur Elon Musk are talking up solar and making significant investments. More importantly, folks like Musk are making the energy revolution appealing by creating products desired by mainstream consumers. I truly applaud these revolutionaries and do whatever I can to support their efforts (as I write this book, my new model S is on order and scheduled for delivery in September 2016).

A good term for products like Tesla's home-battery system, released in 2015 and dubbed the Powerwall, would be *disrupter*. Today every new Tesla product launched in the marketplace gets the same attention that Steve Jobs commanded when announcing new Apple products. *When disruptive products get mainstream attention, a revolution is underway.* Tesla's products are great examples of the big changes in the energy market.

The next breakthrough we see today is the combination of solar energy and storage, which will allow businesses and individuals to be independent, generating their own electricity and storing the electricity they generate on their own. That's where the revolution moves forward a couple notches—and things are really going to get interesting. The old utility model is going to undergo a significant change as a result of the new solar storage offerings. I believe entire homes and businesses are going to become 100 percent self-sufficient in the near future. We've already been involved in making this happen through projects such as the CuisinArt Golf Resort & Spa in the Caribbean, which I'll talk about in an upcoming chapter. If the storage technology improves in the same way we've seen in solar-energy technology in the last few years, by 2023, we're going to see huge disruptions in the utility infrastructure in the United States.

It's definitely exciting to be at the forefront of this revolution. At the same time, it's quite a challenge to be on the ground floor of a revolution, sharing a vision of the future that not many can see. I often think of a quote from Thomas Edison: "It's become appallingly obvious that our technology has exceeded our humanity." Our technology has well exceeded our humanity, and in the US business world, priority is given to the almighty profit.

On the front lines of the energy revolution, a battle is being fought to make solar less expensive and as profitable as the fossil fuels of old so that our humanity can catch up with our technology.

CHAPTER 3
A NEW DIRECTION

The year 2010 was a very exciting one for Inovateus Solar because our business plan had come together. It morphed a bit, and as we became strategic improvisers and learned how to adapt our business plan to the ups and downs—the solar-coaster—of the solar industry, we were able to excel and start to achieve some business success. As we neared the end of that year, we were very excited about the possibilities that lay ahead.

However, a heart-wrenching event threatened to abruptly alter the course of Inovateus Solar.

My father had battled depression and alcoholism throughout his life. And just as I was able to really experience some business success with him, and it seemed our relationship was in its heyday, he unfortunately became ill.

Although he had been through rehabilitation in previous years and had been a good spirits and health for some time, his depression fought back hard. He attempted to control the depression through medication, but he was consulting more than one physician and doing some self-medicating and, unfortunately, two of the medications ultimately pushed back against each other, creating what's known as serotonin syndrome. To combat the syndrome, which basically attempts to shut the body down, he was admitted to the hospital, where he was induced into a coma for ten days. It was hit-and-miss over the Christmas 2010 holiday. Eventually, it was determined that he would survive, but the doctors didn't know what kind of permanent mental or physical damage might have resulted.

He came out of the coma just before New Year's Day 2011, but after that, he wasn't quite the same. He was gloomy and paranoid about the business. At the end of January 2011, his business outlook didn't reflect all the success we had achieved through the end of 2010. There was even talk between my father, George Howard, and my grandfather about closing the company or selling it to a larger corporation or other investors. More than once, I talked them out of it.

At the time, I was executive vice president of the company, and as my father became more detached from the daily business operations, I stepped into more of a leadership role.

For the first five months of 2011, my father continued to battle health problems. Throughout that time, he took twice-weekly hyperbaric oxygen treatments in Chicago to combat the effects of the syndrome, and by May, he seemed to be doing better. He had more energy, his head was clearer, and it looked as if he were making a comeback. Then, on May 17, he was supposed to come to the office for a meeting with me, and he didn't show up. It wasn't unlike him to not be there, but by the end of the business day, concern was beginning to mount.

My mother and I began to look for him, and we notified the police that he was missing. I continued to look for him most of the night. The next morning, the police located his car in the parking lot of a nature conservation club that he had loved as a child and throughout his adult life. They found him outside his car, where he had taken his own life.

While it was obviously a very hard time and a devastating event for my extended family, it was also a hard time for Inovateus Solar. As the oldest son, I took off more than a week from work to take care of my father's affairs.

Employees were uncertain what the future would hold for Inovateus because they knew that my father was the majority shareholder of the company. In my absence, I would later discover, a key employee began a coup of sorts based on a lot of incorrect assumptions. He brought the Inovateus team together and, assuming the company would shut down, told everyone to start looking for work elsewhere.

While I was out of the office, I had a business meeting with the other owners and family members—George Howard; my grandfather, Leonard; and my sister, Ashley (who was a key leader at Logistick)—to decide how to move Inovateus forward. They were prepared to shut the company down if need be; they weren't quite sure of what to do without my father leading the charge. It was the same discussion that had come up a couple of times in the months prior to my father's death, when he was ill.

That's when I felt it was my calling to lead Inovateus Solar. I told George, my grandfather, and my sister that I thought I could lead the company and keep us on our path to success. But it was more than a feeling that I could do it; I felt I *needed* to do it. I cared about the company. In fact, I was more than passionate about it. And I felt very confident about the future of solar and the very essence that had started the company. After some discussion, the other owners supported me in taking the lead.

Never have I listened so closely to myself or felt so confident in a decision that I have made. It's been more than five years since I took the helm, and today I'm even surer about my decision to lead our company into the future and fill the head-coach role left in my father's absence. There is not a day that goes by when I don't miss my father, and I would do anything to have him back. But in the toughest of times, we can become stronger and wiser if we embrace all that is good and focus on the most positive outcomes. That is what I focused on alongside my amazing family.

One reason I felt ready to take the helm at Inovateus was because, over the years, I had been exposed to a number of different leadership styles. My father was a visionary who had the energy to make things happen. He was brave enough to take on challenges and risks and brave enough to try in the face of adversity. He also showed me the power of positive energy and of being a good team player.

Meanwhile, my grandfather had worked on several ventures for a number of years before starting his trucking business Consolidated Services, Inc., a company that had experienced extreme highs and lows. I was very close to my grandfather, so I learned a lot about his leadership style, which was very nurturing. He was also a very well-organized, data-driven person who was good at implementing ideas.

My grandfather on my mother's side, Richard Russell, was a finance guru and served as a controller and CFO for many companies, including a few very large firms such as Eaton. As a kid, I was also very close to Richard (or Papa, as I call him), and having spent many hours with him, I'm sure some of his financial knowledge soaked in.

During my youth, I also looked up to my uncle Charlie Crain, who was president of the western division of GTE (later acquired by Verizon). He was a leader whom I, along with our extended family, really looked up to until his passing in 1994, when he was in his early sixties. Looking back, I can see that as I was growing up, I was surrounded and nurtured by many business leaders, and I guess it's fair to say that it's in my DNA.

I had also learned about leadership at the Fortune 1000 corporation level through my years after college at General Growth Properties. I had worked in the strategic partnerships group under a man named Dave Brown who, through a monthly meeting, did a great job of bringing the team together to make sure we were all on the same page, a practice I brought to Inovateus. I also credit Dave with inspiring one of our core values: engagement. I learned from him the value of engaging the team. While writing this book, I have reached out to Dave and hope to rekindle my relationship with him in the near future.

In addition, I learned from my corporate years the value of a collaborative, flat organization. A pyramid hierarchy, like the one in the much larger company I had worked for, was something I definitely didn't want to see at Inovateus. Tiers, divisions, and levels, I found, hindered a company's ability to tap into creativity, ideas, and innovation and also affected morale.

In discovering more about what makes humans work well together, hierarchical organizations have been found to be quite the opposite of how humans once functioned together in tribes and throughout most of our human existence. As thought leader Simon Sinek discusses extensively in his book *Leaders Eat Last: Why Some Teams Pull Together and Others Don't*, and retired general Stanley McChrystal reiterates in his book *Team of Teams: New Rules of Engagement for a Complex World*, a true leader is a coach and a nurturing soul who brings out the best in people and gets them to bring out the best in themselves.

I see myself as someone who encompasses these characteristics. I'm definitely not the stubborn, dictator-type CEO. For that reason, it can be difficult for some people to understand my ability to effectively lead; their mental model of a CEO is not what makes leaders great.

After I took over as leader of Inovateus, I immediately decided to take the company in a new direction. I wanted to build on the strong foundation that my father, George, and my grandfather had forged. As they had, I wanted to make the world a better place.

In settling my father's estate, we found that the ownership of the businesses (Inovateus and Logistick) was left to my siblings and me. My sister, Ashley Brickley, had until that time, been executive vice president of Logistick, so she took over the role of president of that company. My brother Nick had been an on-air meteorologist in Texas at the time of my father's death and had not been active in either Logistick or Inovateus. By year's end 2011, he and his family had moved back to South Bend to head up research and development at Logistick, an area my father had been very active in. My younger brother, Tyler Kanczuzewski, was in college and interning at Inovateus Solar; after graduation, he joined the company as supply manager. Two younger brothers, Michael and David, are still in school getting the best grades our family has ever seen, as of the writing of this book.

With the bulk of my father's affairs managed for the moment, I returned to the office at Inovateus. I had been gone a little more than a week, but I found a different attitude when I returned. Many people had already started looking for other jobs because they assumed the company was going to shut down. Needless to say, at that point, morale was really low and the door to my father's office had remained closed.

Inovateus and Logistick reside under the same roof, and the mood was similar for Logistick. However, since Inovateus was still relatively new, the mood at that company was grim. It definitely gave me my first challenge, as a leader, to turn the ship around.

I called for a company meeting, during which many people thought I was going to announce a shut down. Instead, I told everyone very much the opposite. I told the team that we were moving forward. We had a plan, I explained, and we were going to be successful. We were going to achieve all the things that we'd wanted to, and we were going to do so in a way that my father would have envisioned. He had believed in Inovateus Solar, and he wanted us to succeed, so I was determined to make his vision reality.

It took a good month of hard work to convince the team that I could do what I had told them I was willing to do and wanted to do. Although I had been overseeing the operations for several months prior to my father's death, I had not yet been viewed as the true leader in charge.

Out of respect for my father and in a reverent manner, I continued to leave his office door closed for a few more weeks. But, ultimately, as part of my efforts to boost morale and to get the team to view me as the company leader, I opened the door to his office and moved some of my belongings into the room, basically setting up camp there. I left a lot of his belongings in the space. I wanted everyone to know that his spirit was still with us. The things that he left us with were still in place; we were just moving ahead. Many of my father's items remain today, including a picture of him standing alongside President George W. Bush, as well as a photo of him and me with President Barack Obama.

President George W. Bush, Tom Kanczuzewski

TJ Kanczuzewski, President Barack Obama, Tom Kanczuzewski

Another item is a statue of a captain at the helm of a ship. That still stands on the windowsill. As long as I work at Inovateus, I will keep that statue on display in honor of our original captain at the helm, starting us on our journey to the uncharted territory of the energy revolution.

By July, morale began to pick up and I had let the "leader of the coup" go.

I started working on a couple of different initiatives, which included redefining our sales approach. We had really achieved quite a bit in 2010 and into 2011. We were able to find our stride, and we discovered that commercial rooftop solar systems were our bread and butter. When we kicked off the company in 2008, we began with a shotgun approach to selling in the market. But after my father's passing, we narrowed our approach and focused our sales team. I worked on improving some of our marketing and Internet presence. And I brought into the company a couple of additional team members who had strong talents in their areas of expertise.

Communications within our team also needed improvement. Our focus was beginning to encompass various functions, such as engineering, project management, and internal operations, and I knew it was important to get everyone on the same page.

We created subgroups, or what I like to call focus groups, so the people who worked closest together could be in strong communication. Those focus groups ultimately included administration, operations, and sales and development.

Our administration team was what most companies would call an operations team. Its members handled our invoicing, cash, accounts receivable, inventory, human resources, and legal and corporate matters. Our operations team essentially handled our engineering, project management, and design—basically all of what I call the "doing" at the company. Our sales-and-development team worked on new opportunities. These team members put proposals and estimates together, and they got in front of customers. Originally, our salespeople were also project managers, but as we continued to grow, it became imperative to have the salespeople more focused on new opportunities and have the project managers become part of the operations team.

Shortly after I took the helm, in an effort to learn more about my own role, I started reaching out to regional CEOs. One of the activities I found very helpful was joining a group called Executive Forums, where I had monthly roundtable discussions with presidents of large companies in our area. I hoped to learn from them, network, and create some new opportunities.

Essentially, I wanted to learn all I could about being the best president I could possibly be. As I learned more and more about leadership from the fall of 2011 into early 2012, I brought that newfound knowledge to the company, which helped to make it even stronger by building on the foundation that my father had been so instrumental in creating.

I also started reaching out to the executives of other companies we were working with, in part to share with them our plan for the future of Inovateus. I needed them to know that we would continue to operate and the company was doing well even though my father had passed so suddenly. I put forth quite a bit of effort to create some strong alliances and trust with our vendors, suppliers, and clients to ensure them that we were going to deliver.

In the winter of 2011, we started to land some larger Fortune 500 accounts, including two large contracts with IKEA to install solar at a couple of the company's distribution centers. In addition to being major wins, those contracts are what actually gave us the foundation to build our team up a bit. As a result of them, we brought in experts in project management and engineering. The engineering professional had expertise in working for utility companies.

Around the same time, the solar market was changing; the price for solar panels dropped incredibly in 2011. That was when mass production of solar panels and other products by Asian manufacturers began to really drive down the cost of solar. When some of our US solar partners started having financial trouble because they couldn't compete with Asian prices, we found ourselves having to develop new manufacturing partners.

The need to constantly improvise in business played very well with my collaborative style of leadership. Being a great collaborator has been really important in getting our company to where it is today and very important in a trailblazing industry such as solar.

As I mentioned, I saw collaboration in action in the business world with my mentor at the Fortune 1000 company. But my collaborative style of leadership also stems, in part, from my musical background. Leading the group at Inovateus in the revolutionary solar industry has led me to tap into the skills and talents I had as a musician. To me, running a business is akin to getting a team of people to play together well. I view myself as something of a conductor overseeing the strategic improvisation of the company. Every member of the team "plays a different instrument," so to speak, and the goal is to get everybody to play together well—to play in harmony. And in business, just as in a band, it's about having rhythm and structure and being in tune.

With those new Fortune 500 accounts that we landed near the end of 2011, we really began growing together as a team at Inovateus. Our employees were reenergized, and we had some new energy on the team, some of which had been infused by the new talents we had hired. The company was heading in a new direction. It was like a new dawn for us. It had been a major blow to lose my father, but in the aftermath, the sun began to shine and we entered a new day as a company.

In January 2012, in addition to defining our focus groups, we began formalizing our future in other ways as well—for instance, by coming up with strategic annual plans. Looking back at the original business plan, we found that we had strayed, largely because of changes in the market, changes in clients' needs, and changes within our team. That really made us see that strategic improvisation was a strength of ours.

With all that we overcame, and all that we improved and changed, by the end of 2012, we had become one of the top twenty solar companies in the United States.

Still, we continued to look to improve our operations. While our administration, operations, sales-and-development structure worked for a while, at the beginning of 2016, we realigned the company during a major growth period. To focus on the four major focus areas of our company—people, execution, cash, and strategy—the teams were aligned for future growth into areas that we now refer to as projects, strategic development, and capital. If you think of a rocket ship propelling itself into space, strategic development is the nose of the ship, capital encompasses the control capsule, and projects are the fuel spewing flames from the base of the rocket.

Our goal is to become focused on projects, products, and ownership of solar assets. Our present-day approach is through the focus of projects (today's fuel) writing our history; capital, which manages the operations in real time; and strategic development, which looks three to five years into the future.

In 2014, prior to this structural change, we also formalized core values that had been fermenting since 2012. Abbreviated as PEACE—which stands for passion, engagement, ambition, creativity, *esprit de corps*—our core values, we decided, were the real reason for our existence. Although we'd been living those core values from the start under my father's leadership, we put them in writing in 2012 and finally "cast them in stone" in 2014 to create a system to guide us into the future.

We have since defined our purpose, our *why* as a company, as "building a brilliant tomorrow." Our core values—PEACE—guide us in delivering on our *why*.

The next five chapters will discuss our five core values and demonstrate how they have advanced our company or led to a positive outcome for a client.

CHAPTER 4

P IS FOR PASSION

There are many different definitions of *passion*. The simplest and the one I like to use is that passion is a very strong feeling about a person or thing. Another way to put it: passion is an intense emotion or a compelling enthusiasm or desire for something.

We started Inovateus Solar with a passion for solar. And I think it's really important for any company to have passion from the start. You must build a team of people who share that passion and believe in the company's mission, which is the real driving force behind the passion.

Let me tell you a story I think conveys what passion for an industry is about.

As I mentioned in chapter 2, in mid-October 2015, I went to the Solar Pioneer Party in Humboldt County, California. The party was organized by Jeff Spies, who has been an integral part of the solar-energy industry for just over ten years as a training specialist and industry advocate. Jeff originally started in the industry working for AEE. Now one of the industry's oldest and largest distributors,

AEE was started by David Katz and cohorts in Redway, California, in 1979. David is an industry visionary with whom Jeff Spies shared a passion for solar. AEE was responsible for fulfilling a need for people in northern California, which was to be a source of electricity in places not reached by the grid. It's interesting to note that some of the early customers and other solar pioneers living off the grid say that they had to grow and sell marijuana to be able to afford solar products, which were very expensive at that time!

My brother Tyler and I went to the invitation-only solar event, which drew a couple of hundred people from across the US. It was really what Jeff called "a noncorporate event," just a celebration of solar energy and a chance to pay tribute to pioneers who helped kick-start the US industry.

Tyler and I knew the event would be largely composed of faces that were new to us, and we didn't go with the intention of making a sale. But we left our families for a weekend and traveled halfway across the country because we were genuinely passionate about solar energy and we wanted to meet and thank some of these pioneers and other industry-leading personalities.

We showed up on Friday afternoon at an older building where AEE was originally founded. (AEE has since moved to a skyscraper in San Francisco and it employs some 1,500 people.) The opening party was just a group of about 150, just hanging out, enjoying refreshments, and talking about solar. Some people in the group had photos of early installations or other gatherings, and a few had brought some solar gadgets with them. The minute Tyler and I walked into the room, we were embraced by these strangers who treated us as if we were old friends.

The next day, we gathered for an all-day event at another facility in Redway, California, where another hundred-plus people joined the group. It was very interesting to see people get together and share stories about how the technology and the industry had changed. For instance, it was amazing to get a clearer picture of how big and clumsy solar panels used to be compared to what they are today. We even shared a few laughs about how the industry had turned from more of a hippie-festival crowd to a corporate culture with suits and ties, which is definitely a very different lifestyle from that started by the solar pioneers.

The one thing that really struck me was that every single person there shared the same passion for solar energy that we have at Inovateus Solar. That passion didn't necessarily equate to commercial or mainstream success in business, but that wasn't the focus for everyone there. Some just felt it was the right thing for environmental or economic reasons; they thought solar energy helped make the world a better place. And it definitely filled the need in northern California to provide electricity to folks who wanted to live off-grid but, at the same time, needed energy.

Solar Pioneer Party

Before Tyler and I left the event on Sunday, we spent some time with David Katz, discussing how he had started his business and talking about Inovateus and where it was at that time. I think David genuinely saw the passion that Tyler and I and the rest of the Inovateus Solar team have. It's the same passion shared by many of those folks we had met at the party back in the 1970s when they were starting their companies.

David Katz and Tyler Kanczuzewski

That passion is one of the main reasons for our success at Inovateus. Obviously, we have to be profitable. We have to manage our business with the right tools, the right leadership team and teammates, and all the things that most businesses operate with. But without a passion for solar and for wanting to make the world a better place, I think we would just be another company trying to make money.

One thing I've found to be a strong guiding force over the years is a message coined by Simon Sinek, who believes companies that are very successful and gain loyal customers and followers are the companies that start with the *why*: Why are we in business? Why are we doing what we do? It's a message that really rings true for Inovateus Solar. As have others who have been successful throughout history, we've succeeded because we started with why we're doing what we're doing.

Sinek uses Martin Luther King as an example of someone who succeeded because he started with the *why*. King felt strongly about racial equality in the United States, and his message on the subject wasn't about what needed to change; it was about his beliefs: they were the same beliefs that others shared, and that's what made his speeches so contagious.

At Inovateus Solar, we started with the *why*, and that *why* was our passion for solar energy. I, along with my father, George Howard, and others wanted to see more of it in the United States. And that definitely helped us to get our business started. But today that passion is just as strong, if not stronger, than when we started.

George expressed that passion in a memorable way when he and I attended a city council meeting in East Lansing as part of our efforts to secure a tax abatement for what will be the largest university solar installation in the US for Michigan State. George moved me when he told the city council, "I was a professor for my entire career and taught about the changes we needed to make as a society to have a healthier environment. When I met Tom Kanczuzewski and TJ Kanczuzewski, I decided it was time to stop professing and make a change. So I put my money where my mouth was because I believed in it."

One great example of Inovateus's passion is the work we've been able to accomplish with Duke Realty Company, a real estate investment trust (REIT) headquartered in Indianapolis, Indiana. As representatives of Inovateus Development, a company of about five people, we first pitched the trust in the fall of 2007. We traveled down to Indianapolis and gave Duke's sustainability team a presentation on solar energy and how the Uni-Solar product could be a good fit for the company. We knew Duke had roof space all across the United States and could be a really good partner for us in finding a home for solar. We presented Duke with the idea of using Uni-Solar's thin film product. It was ideal, we said, for the properties

Duke owned, which included large warehouses, distribution centers, offices, and hospitals.

But we quickly stumbled into some challenges. Since Duke Realty was a REIT, it could not receive the 30 percent tax credit that was available to solar-energy investments at the time. Especially in 2007, that tax-credit ineligibility took away a good chunk of the economic benefits of installing solar.

We left the meeting a bit disappointed because it didn't look as if things would work out. But my father said, "Let's keep at it" because Duke Realty was interested in solar and had asked us to put some proposals together for some of its facilities.

We had a follow-up meeting the next spring. By then, we'd found a few additional solutions, including a way to bring in investment partners who could own the system built on the roofs of Duke buildings while Duke leased its roof space. But the company still wasn't quite ready to move forward; there was some pushback from its executive team. We continued to work closely with Mike Gahimer, who was then the energy manager at Duke. He felt that solar energy could be a good fit for the company.

Finally, when it seemed discussions weren't getting anywhere, we decided to gift Duke Realty a small system so that it could become familiar with the technology. Ultimately, a very small dollar amount was exchanged because the company wouldn't allow us to gift it a system.

We put in a five-kilowatt pilot research-and-development project because we really wanted the company to see the benefits of solar. We used Solyndra technology because Duke had expressed an interest in it. Fortunately, that was our only Solyndra installation because, as I mentioned previously, Solyndra went under shortly after the great panel price drop of 2011. Our desire to get Duke onboard was driven, in part, because we wanted it to offer solar as an option when it was in the construction phase of projects—for example, distribution centers for customers such as Amazon.

Nearly four years later, we still didn't have a contract with Duke, but we maintained a somewhat consistent dialogue to share the happenings of the solar industry. In 2011, we even spent some time trying to find a solar solution for a building the company owned in Goodyear, Arizona, that was being leased by Amazon. Aside from that, we continued without any real business being transacted with Duke.

Then, one day in 2012, we heard about a solar-energy program through Indianapolis Power & Light. The utility company was looking for proposals for solar installations within the Indianapolis area, and it offered what's called a feed-in tariff of $0.20 per kilowatt hour for fifteen years. We got on the phone with Duke and quickly put some proposals together that were accepted and approved by Indianapolis Power & Light. By mid-2014, we had completed ten megawatts of solar installations on Duke Realty buildings in Indianapolis, one four-megawatt installation, and two three-megawatt installations, which together comprised the largest solar-rooftop portfolio in the Midwest at the time. We brought in an investment group to own the systems and sell the electricity to Indianapolis Power & Light, and Duke Realty made money by leasing its roof space. We created value out of space for which the company was getting $0 per square foot.

Duke Realty Installation

Our passion for solar and our desire to see solar work for a company such as Duke Realty kept us moving through that effort for over six years.

Since that kind of passion is what it can take to succeed in this industry, we also look for a real passion for solar when we're looking at new members for the Inovateus team. In our interviews, I ask candidates how they feel they relate to our core values, and I ask them about their backgrounds. Sometimes, I ask them about their lives all the way back to grade school because that lets me know about things they've been passionate about for many years.

We're all wired differently, but the one thing we all can control is how much effort we put into any one thing. At Inovateus, we try to focus on working smarter, not necessarily harder. So we look for people with energy. We look for team members who have shown they've put a lot of effort into things in the past. And a lot of times, we find they've had these traits before they even had their first job. We use a hybrid model of the topgrading technique developed in the 1990s by Brad Smart, which helps us quickly reveal those folks best suited for our company. We find that many folks we don't hire simply didn't prove to us that they had the "passion" gene.

We also look for people who may not have worked in solar energy before. They don't need to know all the technical aspects of solar or even details about the work we do, because we've got great team members who can bring them up to speed. We can teach people the technical aspects of solar, but we can't teach them how to be passionate.

Internally, passion is the catalyst behind much of what we do. Whether we're selling solar products, designing installations, engineering a system, project managing, or even constructing a project, the passion is the flame, the driver, the mover. And we are always drawn to customers who share the same passion for solar or a passion for the industry they work in.

At Inovateus, we do many things we've never done before. In many cases, the projects are firsts for our customers as well. So our management team members look for the passion in each other and in themselves, and we try to harness that passion and channel it into our systems and processes. From the initial phone call with a client through the proposal, preliminary designs, site visits, analysis, calculations of how much electricity could be produced—every step of the process requires a true passion on the part of team members.

We also bring that passion to team activities. We often start off meetings by asking, "What was your personal best from the previous week?" It's not necessarily business related. It may be family, a sport they play, or an activity they're involved in. That sharing helps us learn more about each other, brings our team together, and makes us stronger.

Every other day, I hear stories from team members who have been asked by friends or family about solar and the solar industry, why they work at Inovateus, and what they do. Is this industry going to survive? Is it going to last? Is it a bubble? Is Inovateus Solar a secure place to work? Is it a secure future? Some of these are hard questions that people want answers to.

So our team members have to be passionate about solar since we work in an industry that's still gaining a foothold. They have to be passionate about more than just a paycheck. They must have a passion for excellence and for being part of this group. That passion I felt when I met with my father and George Howard—that's the same passion our team needs in order to be successful and for the company to continue its forward progress.

CHAPTER 5

E IS FOR ENGAGE

In this industry, it's important to be as engaging as possible.

Engagement is a core value that Inovateus had from the beginning. It was actually the one core value that other people were trying to describe to us when they talked about why they liked working with us, but it took a while to define it, to come up with just the right word.

Soon after we kicked off Inovateus Solar LLC in 2008, we began making presentations to companies and setting up relationships with vendors. Some of the feedback we were getting from people then was that they really liked working with us. They found us to be genuine and fun, the kind of people they felt comfortable being acquainted with beyond the working day. I remember a customer in New Jersey, in 2010, telling us how different we were from people on the East Coast, where it was a "fast-paced, in-your-face mentality." "You guys are, like, freaky nice," he told me. "It's almost uncomfortable at times how nice you guys are."

Obviously, he was joking—or maybe not—but we began to realize that our attitude was a strength. Still, we couldn't quite put our finger on a word to really describe this concept we were hearing from the people we worked with.

One day, Brian Lynch, who now works for SolarWorld, said, "You guys encompass the Hoosier values." When I asked him to explain, he repeated what others were saying: we were nice, but also, we had Midwestern values, which differentiated us from most people in the solar industry. The West Coast, he explained, where most of the people in the solar industry live and work, has a progressive but sometimes edgy and a little materialistic lifestyle. A smaller percentage of solar people were on the East Coast, and they were very matter-of-fact, putting their feelings out there and not holding back, for better or for worse.

After scouring a dictionary and thesaurus, I came across the terms *engagement* and the *act of engaging*. One definition read, "to pledge oneself." I realized that was pretty close to what they were talking about. We are engaging. We talk to people, we educate people, we get out of the office and visit clients, and we invite them to our office. We go out of our way for customers, we go out of our way to teach people about solar, we give presentations at colleges and high schools, and we bring grade-school students in to learn how solar energy works. We go to industry trade shows and meet with vendors. And we get involved in after-hours activities including attending football games, basketball games, and so on.

We engage with customers, vendors, and with each other, and when we do so, we have a high level of communication. That's important in an industry in which many customers are unfamiliar with the product. They don't understand exactly how solar works, and it's up to us to explain it to them throughout their projects.

This attention to engagement and education is atypical of most construction projects. People have been building houses and skyscrapers for years, but they don't necessarily question the way in which such projects are undertaken. They don't need to know all the ins and outs of how a structure is going to be erected, and they often don't spend a lot of time wanting to know all the details as a project progresses.

But because solar is an emerging industry, customers usually have a lot of questions and a genuinely elevated level of interest. So we need to be engaging in order to pique and continue to hold their interest throughout a project, from concept to reality. And that involves a high level of communication at all times.

We've discovered this to be true because of those few times when we haven't engaged as much as possible. The hurdles, we've found, arise when we haven't engaged with each other at the company and we haven't communicated at a consistently high level to ensure that information flows to everyone involved in a project. It's a challenge to always be engaging, but we work at it every day because we see the value of it in every aspect of our work.

Probably the project that best demonstrates how we engage is the CuisinArt Golf Resort & Spa on the Caribbean island of Anguilla. Our level of engagement truly led to the success of this project.

The resort is owned by Leandro (Lee) Rizzuto, the billionaire CEO of the Conair Corporation, which acquired the CuisinArt kitchen appliance company in 1989. The resort was Lee Rizzuto's first venture into the hotel business, started when his dreams of building a lavish home for entertaining clients, friends, and family were dashed by a local law in Anguilla. After he purchased the property on Rendezvous Bay, Lee discovered that beachfront construction by foreigners was restricted to the building of resort properties only. Undeterred, he expanded his entrepreneurial portfolio to the hospitality industry.

From the start, the project experienced its share of challenges, including a hurricane that caused structural damage a week before the facility was to open. But ultimately, it opened as a luxury resort complete with world-class chefs, restaurants, and a spa, all of which use Conair and CuisinArt products (naturally!).

While the resort has been a success, attracting travelers from around the world, a new challenge emerged after it opened. Powering the energy-intensive property became a significant expense because of the cost of electricity, which is generated at the island's power plant from diesel fuel that is hauled in from Venezuela or Barbados—a very costly resource. At the writing of this book, electricity on the island costs around $0.45 a kilowatt-hour (compared to the $0.10 a kilowatt-hour for electricity in South Bend, Indiana). In short, island pricing for electricity is four and a half times what it is in the American Midwest. So while CuisinArt is a magnificent resort offering guests the ultimate in luxury and modern conveniences—air conditioning, televisions, modern appliances, restaurants, and fresh linens (cleaned in the on-site laundry)—those comforts come with a hefty price tag.

A few years after it opened, the resort added the island's only golf course to its amenities. The Greg Norman Signature Golf Course was part of a nearby development that fell apart in the wake of the 2008 economic collapse. Lee worked out a deal with the Anguilla government to take over the golf course and the development that was already underway, which included condominiums and apartments.

Now here's the catch: When you own a golf course in Anguilla, you have to water it a couple of times a day because the island has an arid climate. While Anguilla is surrounded by ocean, all that saltwater does nothing to create lush golf greens. So fresh water for the course has to be made through reverse osmosis, which requires the installation of a plant next to the course. And what does a reverse-osmosis plant need for it to operate? Electricity, of course. That added to the costs the resort was already shouldering.

To address the situation, Lee hired a chief engineer, Rory Purcell, who decided solar might be a good option to help reduce the costs of electricity for the island facilities. They immediately considered how solar would help generate electricity when the sun was out and how that electricity could also be stored to use when the sun didn't shine.

Lee contacted Peter Foss, a good friend of his from the US energy conglomerate General Electric (GE), and told him about the situation at the CuisinArt Resort. Coincidentally, Inovateus had just finished a solar-energy carport that featured GE's latest electric car charging stations at the GE Industrial Solutions headquarters in Plainville, Connecticut. We were also working on some other initiatives with GE, which involved trying to use some of the company's products.

In our work with the corporate giant, we found many of the divisions to be siloed. So we arranged a number of meetings in which we introduced GE people from the company's different divisions to each other in an effort to find ways to marry up the company's products. One of those meetings was in Schenectady, New York, where GE's solar-energy research team is located.

So when Lee approached GE about solar, the company brought us in on the project. One of our account executives, Peter Rienks, and I met with Jeff Immelt, GE's CEO, in Chicago to talk about it. At that meeting, Jeff expressed his belief that the project could be a game changer. "It's up to you guys. Make this happen; make it work," he said.

We went back to the Inovateus offices and continued to engage with GE product leaders, with Rory Purcell at CuisinArt, and with our own team members to try to figure out how we were going to make the project a reality. In no time at all we realized it was going to be a difficult project, like nothing we had done before. In fact, it was the first project of its kind. Many of the manufacturers we were talking to hadn't done anything like it, and a lot of the products they had weren't capable of connecting with the other products we needed to make it happen.

But we kept engaged with Rory, who continued to share his vision—and Lee's vision—with us. In December 2013, Joe Jancauskas, who was our vice president of engineering at the time, traveled with me to the CuisinArt Resort to meet with Lee and Rory and other key managers who oversaw the property. We all pieced together how we thought the system could work.

In that meeting, Lee explained how the resort was going to double in size, adding new condos and a tower as part of the deal he had made with the government when he took over the golf course and the development that had faltered in the recession. He also talked about some of his other business ventures, and I remember what an impact he had on me. Here was a true visionary, a businessperson who had done things that hadn't been done before, and who had the resources to get the ideas at the resort done. I realized what a huge opportunity Inovateus had before it: we had a partner who was willing to see a unique and challenging project through to its completion.

On our return to Indiana, Joe and I really dug into the project, getting together the group we needed to make this solar installation a reality.

In truth, a few members of the Inovateus team weren't fully engaging in the process. One of the ways in which they disengaged was to withhold information we needed to move the project forward.

A few of those folks are no longer with us because they were trying to stop us from doing what we wanted to do with this customer. Their level of engagement wasn't where we needed it to be to succeed. But through that inner turmoil, we continued to engage and move forward by keeping a high level of communication with our customer and the manufacturers.

At the end of the day, we used very few GE products on the project. Even though the company is a world leader in technology, it still had not brought some of the products to market that we needed. However, Lee Rizzuto wanted to move forward. He wanted to achieve the initiative, even if GE didn't have the products. So we got together with around ten different manufacturers and pieced together such an incredible system that even years later, I'm still amazed by it.

Throughout the entire project, we had to stay engaged with people in numerous locations, from our base in Indiana to numerous manufacturers around the country to the folks in Anguilla (which is a long journey—but a trip I didn't mind making when it worked out!). We also had to engage with an engineering company on the West Coast to help expand our capabilities. And we had to engage with Anguilla Electricity Company Limited (ANGLEC).

Engaging with ANGLEC was a challenge because the company did not think it was possible for us to do this project and because there was some misunderstanding of the project itself.

We originally obtained all the needed approvals and permits from the island's planning authority in order for the project to move ahead to completion. However, when commissioning the system—basically flipping the switch to turn it on for the first time—we asked the utility company to disconnect the power, to disconnect the resort from the grid. Everything went fine and the system worked as planned. But the next day, when we asked the utility company to turn the power back on, it refused to do so because it didn't understand the project. So we had to meet with the utility company officials again and explain to them what we had built and so on. In the end, we got our message across, and the utility company turned the power back on.

When we built the project in Anguilla, we used local labor, creating jobs on the island with a team of people who had never built a similar project. We had to engage with that local team to teach them how to do what needed to be done. We managed the entire process and hired a former general manager of the utility company, Tommy Hodge, to act as one of our on-site project managers. Tommy was extremely instrumental in directing the labor and installation of the project, and his role was critical when we had to explain the installation to the local utility after running into difficulty after the solar was commissioned.

Anguilla CuisinArt Installation

As of December 2015, the utility still had not reconnected the power to the reverse-osmosis plant that served the golf course, where we had installed a one-megawatt, off-grid solar system. We believe, now, that it had not been reconnected because we had taken that plant—one of ANGLEC's largest customers—off the grid, which was undoubtedly a fairly significant economic hit to the utility.

Today, as a way of retaining more control over the island's electricity, ANGLEC has been instrumental in passing a law against connecting solar systems to its grid. And we haven't been able to get the utility company to engage with us at the level it did in the past.

But this Anguilla installation is a perfect example of what the solar revolution faces every day. Half of the people in Anguilla can't afford electricity because it's so expensive, yet the utility company there won't allow people to install solar.

In fact, CuisinArt is making its own water supply independent of the electric utility. It's actually creating more water than it needs and, consequently, sells water back to others on the island who need it for drinking, restaurants, homes, and businesses.

CHAPTER 6

A IS FOR AMBITION

As I've mentioned, when Inovateus Solar LLC was launched, we began to pitch solar energy to companies and utilities at the Fortune 500 level. We were a small team (five people), and we didn't have a very strong portfolio at the time. All we had was a plan. Yet, at every meeting we had, we were able to command the attention of prospective customers. Even though we didn't have a track record and hadn't accomplished much of anything that we put in our business plan, our customers came to trust us and believe in us.

The reason I think they trusted and believed in us was that we were very ambitious in what we wanted to accomplish. That's another of our core values: ambition. We started from nothing, and today we're in the top twenty of solar companies, and that takes the kind of ambition we've shown. We have a goal of moving our company to the top five in the solar industry by the beginning of 2023, and if we're going to achieve that, it's going to take even more ambition.

But when you're very focused on something in the way we were and continue to be, it definitely comes through to others. As I mentioned earlier, we started with our *why* (as Simon Sinek calls it): We knew we wanted to create an amazing tomorrow through energy independence. And we knew it would take ambition to get us there.

The ambition we have at Inovateus started with my father, who was always looking for new ideas. George Howard, who joined my father at Inovateus, was also very ambitious. He had developed and headed up a department at Notre Dame and led a national association. I feel I'm ambitious as well, but I definitely benefitted from having mentors who were driven to see Inovateus create an amazing future for solar.

So, starting from just a handful of people who had no real track record and who pitched to much larger corporations and utilities, we were able to create relationships we nurtured over time. We made contact with companies such as Procter & Gamble, American Electric Power Company, and Duke Realty, and we kept them informed about what we were doing. We checked in every six months or so, and we showed them that we were serious about solar.

Indiana Michigan Power was one of the companies we pitched in 2008, in our earlier years. The utility serves the north-central and northeastern part of Indiana, along with southwest Michigan. Indiana Michigan Power is a wholly owned subsidiary of American Electric Power (AEP), a large, publicly traded Fortune 500 company.

When we pitched Indiana Michigan Power on solar, it was definitely a one-sided meeting in which we described to the utility executives the benefits, progress, and higher efficiencies coming about in the solar industry. During the meeting, the utility executives' answer to our enthusiasm was pretty quick and to the point: "It's very nice to meet you. Thanks for letting us know what's happening in solar energy. However, solar energy is too expensive for us to invest in, and we don't see solar being a viable part of our energy portfolio in the near or distant future." In other words, we operated in their service territory, so they gave us the courtesy of their time, but they had no intention of installing solar.

A few years later, near the end of 2011, Indiana Michigan Power was facing a rate case. The company was looking for around $300 million from its customers, and it had to appear before the Indiana Utilities Regulatory Commission (IURC) to justify the rate increase.

Now, we've worked closely with our state government officials in the past, and a few of them have been progressive and tried to advance renewable energy, including solar energy. Through those connections, we were introduced to a lobby group that was trying to push solar forward, and they knew we had had conversations with Indiana Michigan Power in the past. They reached out to us because they thought this rate case could be an opportunity for us to get in front of Indiana Michigan Power again but, this time, down in Indianapolis at the state capitol building with the participation of the Indiana Utilities Regulatory Commission. So we took up the offer from the lobby group and we intervened in the Indiana Michigan Power rate case.

During the rate case, we learned that in a few years down the road, Indiana Michigan Power would have to upgrade its Cook Nuclear Power Plant because the plant was nearing the end of its thirty-year service life. That would cost the rate payers another $1.4 billion.

We took the stand that solar was becoming cheaper, the technology had improved, the efficiencies were better, and Indiana Michigan Power should spend its money to work with customers who wanted solar energy. The utility was opposed to our position at the time, but within the procedural framework of the rate case, we were given time to talk at the podium, which gave us the chance to share information about ourselves and the solar industry with members of the utility company and the regulatory commission.

During the case proceedings, when we testified that we thought the power company's data was about five years behind the times, the regulatory commission requested that Indiana Michigan Power conduct more up-to-date research on solar energy. The IURC agreed with us that the utility company's data was lacking and even went so far as to suggest the power company might be deliberately using older data because it was strongly opposed to solar.

In spite of our efforts, we didn't prevail in the rate case. Indiana Michigan Power wasn't required to implement a solar-energy project or a solar-installation program for its customers.

However, during one of those meetings, I was able to sit next to Paul Chodak, president of Indiana Michigan Power. It's probably an unlikely scenario that, during a rate case, you make friends with someone on what might be considered the opposing team, but I was able to create a relationship with the leader of that organization.

I shared with Paul our goals, visions, and ambitions at Inovateus Solar. I told him that we were planning to be a top solar-energy company in the years to come. "Paul, we'd really like to work with you guys," I told him. "We'd like to help you adapt to a changing world where renewable energy is becoming more prevalent. As your customers ask about solar energy, wouldn't it be great if Indiana Michigan Power could provide them some answers and some expertise?"

I reminded him that we were right in the power company's backyard and in its service territory. "It makes sense to work with us," I told him. "The reason we're involved in this rate case is so that we can have the stage to share more information with you." I asked if we could check in with the company periodically and maybe some day work with Indiana Michigan Power, and he agreed to let me keep him up to speed through periodic updates and occasional lunch meetings.

Paul had seen our work with the solar-energy carport at General Electric's industrial headquarters in Connecticut, and Indiana Michigan Power had a car charging station program for its customers. I was able to meet with Paul to discuss the solar car charging stations, and I was also able to report back to him about many of the projects we had finished. One of those I discussed with him was a five-megawatt rooftop system for IKEA in Perryville, Maryland, which we finished in 2013. (It's still one of the largest rooftop solar systems in the country.) I'll expand on this project in chapter 8.

For three years, I met with Paul to talk about Inovateus. In those meetings, Paul would ask me how things were going at Inovateus Solar, and he asked me about the types of projects we were doing and whether we were able to achieve some of the goals that we had originally set out to achieve. Essentially, I believe what he was asking was whether our ambitions were genuine. Were we able to achieve things that we had been ambitious about?

Then, in 2014, Indiana Michigan Power announced that it was going to start a solar-energy pilot project for sixteen megawatts at five different locations within its service territory. We were able to be one of the first companies invited to bid on the Indiana Michigan Power solar pilot request for proposals (RFP).

The bid was facilitated by the parent company, AEP, headquartered in Columbus, Ohio. We learned a lot about large utility business models and the fact that Indiana Michigan Power didn't actually generate its own electricity. It was AEP, the parent company, that would own these new solar systems.

Deer Creek Project for Indiana Michigan Power

We won the first solar pilot project that Indiana Michigan Power had, a project of two and a half megawatts in Marion, Indiana, just south of Fort Wayne, Indiana. It's 95 percent complete as I write this book, and it's our largest project in the South Bend area.

At the end of 2014, Paul Chodak announced that Indiana Michigan Power was looking to bring somewhere between eight hundred megawatts and one gigawatt of solar energy into its portfolio in the next twenty years. That is a monumental shift for Indiana Michigan Power, compared to where the company was back in 2008.

So our ambition to have Inovateus be a top-five solar company might seem unrealistic to companies that are currently within the top five and are based on the West Coast or the East Coast. But if utilities in the Midwest bring on solar-energy programs like the ones being planned by Indiana Michigan Power, then we are well positioned to take our place at the top of the heap.

Late in the third quarter of 2015, DTE Energy put out an RFP for up to sixty megawatts of solar energy within its service territories. We had already completed a couple of smaller projects with the company and were finishing up a five hundred-kilowatt project in Brownstown, Michigan. We had previously explained to the company's leaders that we wanted to be the go-to solar company for any installations in the future.

With the announcement of the sixty-megawatt project, we put together a strong proposal for multiple sites. But the company was looking mostly for solar farms, which are large, ground-mounted systems. At about the middle of the fourth quarter of the year, we were notified by DTE that we were shortlisted for the sixty-megawatt RFP. So, for the rest of 2015, John Jackson, our executive vice president, who oversees operations, and Peter Rienks, our senior account executive, began dedicating nearly all their time to these projects while other members of the team focused on preliminary designs and the engineering requirements. Our work paid off. Early in November 2015, Inovateus signed a contract with DTE for a sixty-megawatt solar project comprising three different sites: two in Lapeer, Michigan, totaling fifty-five megawatts, and one five-megawatt installation in downtown Detroit. The sixty-megawatt solar project will be the largest system in the Midwest.

Lapeer, Michigan Project for DTE

This is a huge milestone for solar energy in the Midwest. To date, our largest project was the installation of a twenty-megawatt system in Rincon, Georgia, for Georgia Power. The DTE project is three times that size and is propelling us into the top twenty solar installers in the nation.

Rincon, Georgia

Needless to say, taking on a project the size of the DTE sixty-watter is ambitious. From where we started to today, our ambition has fueled what we've been able to accomplish. Ambition is "the guts" of our core values. It takes a lot of guts to take on these challenges, but if we keep at it, we'll reach our big hairy audacious goal (BHAG, as coined by the author Jim Collins) of being a top-five solar-energy company in the United States by the end of 2022.

As a leader, I have to be ambitious and push our team forward. But that involves motivating people to drive them to succeed without driving them crazy! The management team must also show our other team members that we believe in them too.

In leading this company, I see my main role as being the visionary. I'm a big-picture person. I can see the future. I can touch it, taste it, see it, feel it in my mind. I can paint the picture in my head. I can even paint it for others. My visions are very ambitious, so ambitious that, sometimes, when I share the vision of where Inovateus Solar can go in the future, people give me a puzzled look. They don't quite understand how we will get there. But I've been blessed with having a very skilled team to help fill in the pieces of the foundation to get where we want to go. I'm only an effective visionary if I have the necessary supporting cast to accomplish what I believe we can accomplish.

For instance, John Jackson is an implementer, a detail person. Often, when I share a vision with him, he sends me a chart the next day that shows the vision on paper. Then there's Lindsey Bauer, our vice president of finance and administration, who puts together the budget, manages the cash flow, and builds an administration team that helps us do what we do day in and day out. And Tom Brown, our vice president of sales and development, manages our sales team and keeps the opportunities coming. We've also built a team of industry leaders in solar-energy engineering and design. We've aligned ourselves with the best manufacturers in the industry. And our team is constantly doing research and development to find the best products we need.

When we hire new members for the team, we look for people with a variety of backgrounds. But we also look for people who have ambition and aren't afraid to be trailblazers because, as I've said, a lot of what we do nobody has done before. We need people who genuinely want to do big things. When we interview, I believe we rank ambition higher than experience.

So as an ambitious leader, I have to take care to ensure that the vision for Inovateus Solar is clearly communicated. I must also ensure that I continue to have a team that can articulate that vision because it can get a little confusing with all the emerging technologies and information. To do that, I must continually understand my strengths and weaknesses, which means I must spend almost as much time focused on my weaknesses as my strengths.

My weakness lies in planning details. Visionaries are not usually very analytical or detail-minded people. That definitely holds true for me. I'm not good with the minutiae. For example, even though I wrote the original business plan in 2008, we must constantly update the plan, and it takes a whole team to vet all of the details. I can understand where we're going and how we're going to get there, but I've found that I need a very strong team to help lay the foundation.

Being ambitious is a great thing, but it's also important to be conscious of the challenges that strong ambitions can bring about. Ambition will expose both strengths and weaknesses, and I've found that the weaknesses must be embraced if the challenges are to be overcome.

CHAPTER 7

C IS FOR CREATIVITY

As we've grown, we've found ourselves becoming more and more creative, and this is true for me too, as a leader.

Creativity is something I've embraced from an early age. When I was very young, I used to spend time with my maternal grandfather, and together, we'd listen to Garrison Keillor's *A Prairie Home Companion* radio show. I was a big fan of the show because of all the different characters and the improvisation. I really enjoyed the mental pictures that show painted in my mind. I was such a fan that one day I dressed up as Garrison Keillor, complete with bowtie and suspenders, and I sent the picture to the show's host. Sure enough, a couple of weeks later, I got a postcard from Garrison Keillor himself, thanking me for the picture and expressing delight and surprise at having such a young fan.

About a year later, I found out that I could play piano by ear, so I started taking lessons and I learned to read music. My love for music grew throughout my school years, and my talent with instruments such as piano, guitar, bass guitar, and others started to blossom. I joined forces with some of my friends in high school and we created a band called Funk Harmony Park.

We wrote a lot of our own music and we recorded an album of songs that contained lyrics somewhat like folk tales. I think that traces back to my love and passion for Garrison Keillor and his improvisation and storytelling. A lot of the songs we performed were improvised, and we mimicked many of the rock and jazz and blues artists of the day.

For a time, we actually had a following and performed around the region. We played in venues for young adults of twenty-one years of age and older (I was only eighteen). When we played in South Bend, our friends couldn't come to our gigs because they were still underage for the venue. Then, the members of the group went off to different colleges. I moved to Chicago and went to Columbia College to study music under William Russo, an accomplished composer and jazz artist who was the head of the music department at that time.

Today the members of the band—Nick Kovach, George Bonin, Mark Criniti, Connor O'Sullivan, and I—remain very close. We get together once a year for a large benefit concert in South Bend around Christmas. One hundred percent of the proceeds go to Big Brothers Big Sisters.

Now, as I've said, leading Inovateus Solar is something like playing in a band. We have a team of people who have many different skills and talents, and when we bring those talents together for the good of a customer and a project, it's a harmonious thing. In fact, many people in the solar industry consider Inovateus Solar to be one of the rock stars of the solar-energy industry.

I think it ties into our creativity. There are some very clear demonstrations of that creativity in our team activities. One of the more prominent examples is the Solar Battle of the Bands, an after-party event at the Intersolar North America conference, which is held the second week of July in San Francisco. A limited number of bands are selected by audition to play for an invitation-only crowd at the event.

A few years ago, when we heard about the event, my brother Tyler and I decided it was right up Inovateus Solar's alley. In early 2013, we assembled a band among our employees and submitted our application to Quick Mount PV, the event organizer. We were accepted as one of the bands to play, and that year, at the conference, our band took the stage, playing against musical groups from some of the top companies in the industry including SolarCity, Sungevity, and SMA Solar Technology AG.

When we took the stage to play before some 1,400 attendees, our logo was displayed on the wall behind us. Dressed up as members of the Beatles' fictional Sgt. Pepper's Lonely Hearts Club Band, we played a twenty-minute set of mixed songs we thought related to the solar-energy industry. We started off with "Taking Care of Business" by Bachman Turner Overdrive, and we played "Fire" by Jimmy Hendrix, "Jumpin' Jack Flash" by the Rolling Stones, and "Bust a Move" by Young MC. We were well received. In fact, we took second place that year.

When we finished playing, a number of people came up to shake hands with us and tell us they liked our music, but most of them had never heard of Inovateus Solar, and they wanted to hear more about the company. So we shared our mission, goals, and passion with everyone who was interested. People found it very interesting that we were from the Midwest, because most of the companies (and bands) at the Solar Battle of the Bands were from California—many of them had headquarters in the San Francisco area.

The whole event kind of blew me away because I was playing and singing with our company band, but I'm also president of the company. Plus, the event helped us make more of a name for ourselves. We even made some special business cards for the group that went to the tradeshow. They included each person's name and title, along with the instrument that person played in the band. For instance, mine read, "TJ Kanczuzewski, President, Keyboards, Vocals." People really got a kick out of those cards and told us how creative they thought the idea was.

We've played in the Solar Battle of the Bands ever since, which has created something of a fan base for us and is really a testament to the impact the activity can have. In fact, we've made new contacts and built solar projects based on meeting various company representatives at the Solar Battle of the Bands.

Our creativity with music has actually equated to business for us. For example, in the summer of 2015, Tom Brown met with a group in El Salvador that was trying to initiate some solar-related engineering with us in that country. At the meeting, Tom introduced himself and handed his business card to the engineer from El Salvador. When the engineer saw the logo, he remembered our band, Reverend Ray and the Everlasting Incentive, and he told Tom, "I love Reverend Ray. You're one of my favorite bands."

Now, being in a band really has nothing to do with solar energy, other than the fact that it demonstrates our creativity. It just goes to show that anything we put our minds to, anything we put our creative juices to, can be successful. We can even have a fan thousands of miles away in another country.

The fact that we can create a band within our company lets our customers know that we can, essentially, create all different types of projects with them. We can include them in our band.

And as we continue to grow, we must also grow as a team. That means bringing on new team members—new players in the band, if you will—who bring with them other talents we don't currently have in order to provide more capabilities for our customers.

As our company has grown, we've continued to take on more projects, more customers, and new challenges. Every project that we work on with our customers conveys a sense of our creativity. One that comes to mind in particular was for the solar-energy charging stations that we built for General Electric in Connecticut. I mentioned this earlier as the project that, ultimately, led to our relationship with GE and the CuisinArt Resort project in Anguilla.

That project was part of GE's launch of electric car charging stations across the globe. Since it was a somewhat unique technology, it required some creativity on our part to figure out how to tie those electric car charging stations into solar-powered stations. Since then, we've built more electric car charging stations for GE and landed a very large solar project with Michigan State University that will be one of the largest university solar installations and the largest university solar charging station project in the country. One of the really creative aspects of these stations is that we've set them up as branding opportunities by decorating them with the client's logos.

So creativity, as it applies to Inovateus, is about everything from forming a company band to helping get our name out there, to being strategic improvisers to ensure that every project makes the most of a client's brand.

CHAPTER 8

E IS FOR ESPRIT DE CORPS (TEAM SPIRIT)

There are several meanings for the term *esprit de corps*, but the meaning I take away is "team spirit." If there's some magic in the world, it's the magic that happens when people combine their talents and skills toward a common goal.

When we were formulating this core value, Tom Brown shared with me many examples of the teamwork concept that they studied and implemented in the armed forces (Tom had served in the US Marine Corps). I was intrigued by the teamwork concept and knew we had to embrace the same idea as a business when we're working together on solar-energy opportunities.

I'm also a big fan of teamwork in sports. I love watching professional football and basketball, and being from South Bend, Indiana, I've naturally been a big fan of the University of Notre Dame, going all the way back to the days when Coach Lou Holtz led the team. Coach Holtz was a great motivator. He had a good strategy, and he talked about teamwork and about the mission and the core values of the University of Notre Dame. As a kid, I saw him get involved in a lot of activities in the area. In fact, watching him work on and off the field, I consider him a mentor of mine.

Left to right: John Jackson, Lou Holtz, Lindsey Foley, TJ Kanczuzewski and Tom Brown

Running a business is a lot like coaching a sports team, although I've found that many teams are run better than businesses. One of the major concepts of teamwork is getting everyone to work well together; everyone has to be in sync to get the results needed to be successful.

Even superstars of the game, such as Michael Jordan, had a strong team—the Chicago Bulls—surrounding him. Jordan may have been a great player in his own right, but he was also a great team player and a great inspirational leader to his team. I personally credit much of Michael's success to his tremendous coach, Phil Jackson, who is one of the rare coaches/leaders in the modern world who understands the transcending magic of teamwork and being part of a unified entity. I love Phil Jackson's book, *Eleven Rings*, coauthored with Hugh Delehanty. In it, he describes teamwork on the basketball court and how the greatest team player on the court at any given time is the one most aware of the other nine players. I really think this holds true for our company's teamwork, and at any given time, our best team player is the one who is most aware of what his coworkers and company is focused on.

When we started Inovateus Solar, we had a strong sense of team spirit. But as we grew and brought more people onboard, I think we paid less attention to instilling that spirit in new members of the team. A few years after Inovateus had been in operation, I recognized that some of the hurdles and challenges we faced were due to a lack of team mentality. Our systems and processes were slowed down by what we thought were bottlenecks in the way we did things. But when we examined where the problems truly were, we found they weren't in the actual systems or processes but, rather, the players involved. People didn't work together as well as they could or should.

That frustrated me because I believe so firmly in the team concept. It was a concept I had really learned firsthand from my father, who pointed it out to me when I was playing basketball in eighth grade. I really wanted to be in the starting line-up with the team, but at the beginning of the year I was struggling. I thought it was because I wasn't making my shots or wasn't playing a good enough defense. But my father made me realize that the problem was because I was too focused on myself and not enough on the team.

Worse still, when I wasn't doing well individually, I started to create negative energy, which was worse for the team overall. So my father really helped me realize the importance of teamwork. Once I focused on that instead of focusing only on making my shots, it became a matter of looking out for others. For example, once I got the ball, it was about helping the best shooters get in position to make a shot and score for our team. If I were on defense, it would be about keeping my opponent from scoring and also about working with the other defenders on my team to devise a strategy to stop players on the opposing team from scoring points.

At Inovateus, the best example of what we mean by *esprit de corps* is a five-megawatt project for IKEA which eventually turned into many more projects with IKEA.

In 2011 and into 2012, an opportunity came to us from Uni-Solar, which made lightweight, flexible solar panels. Uni-Solar had already completed some conversations with IKEA.

At that time, IKEA had a global solar-energy plan and wanted all of its buildings to have solar-energy systems. The company has two large distribution centers on the East Coast, one in Perryville, Maryland, and the other in Westampton, New Jersey. The buildings are large—so big, in fact, that the roofs are made of thermoplastic polyolefin (TPO), a single-ply, roofing membrane that can only withstand a certain amount of weight on top of it.

IKEA in Perryville, Maryland

The larger of the two buildings, in Perryville, Maryland, encompasses 1.8 million square feet. It sits on roughly forty acres near Chesapeake Bay. Cargo ships bringing IKEA products from Europe drop off their shipments in the port in Chesapeake Bay. Those items are then transported to the IKEA distribution center from where they are shipped out to IKEA retail stores throughout the country.

Since the building is so big and the roof can only withstand a certain number of pounds per square foot, Uni-Solar was a good solar product to use; the panels are lightweight and flexible, and they peel and stick right onto the roofing material.

We had been selling and installing Uni-Solar's products for some time, so Uni-Solar's John Williams recommended us to IKEA. But IKEA still wanted us to go through a request-for-proposal (RFP) process because, at the time, Uni-Solar had three companies approved to install solar for it in the United States. The Perryville project was originally planned to be around four megawatts, and the Westampton project was going to be two and a half megawatts. We went through the RFP process, met with the IKEA corporate team, put together a proposal, and ultimately won the bid.

However, as we were putting our bid package together, team members expressed varying opinions on how we should build the project and if we should even attempt to do it at all.

Those who wanted us to undertake the project included Tom Brown—at that time, a senior account executive—who thought that working with IKEA would be a great opportunity for Inovateus. He believed in our capabilities to design, install, and deliver the project to IKEA, regardless of its size and complexity.

However, our project-management team and the engineering and design team had concerns. From a project-management standpoint, it seemed to be a huge undertaking. In order to install the solar panels, another layer of roofing had to first be installed on top of the TPO. This was a preventive measure to accommodate any potential problems with the roof or the panels. Obviously, Inovateus is not a roofing company, but this project involved a lot of roofing construction, along with other things we needed to learn on the fly.

So there was a bit of turmoil inside the company, and things got a little heated. That's when I realized that the major obstacle with the project was that we weren't working as a team.

To combat the situation and to turn things around, I called a meeting with approximately ten people—in sales, engineering, design, project management, and other departments—to talk about the project. I now refer to the type of discussion we had that day as "a brutally honest conversation." A couple of years later, I read Pat Lencioni's book *The Five Dysfunctions of a Team*, which I highly recommend to any businessperson. It made me realize we were dealing with some of what he had written about, and our solution of holding a face-to-face meeting—a brutally honest meeting—was one of his concepts.

Earlier, I said there's magic when people work toward a common goal. Well, some sort of magic also happens when you get people together in person. With so much technology today, people can go a long time without ever talking face-to-face. When I sat everybody down together, all of a sudden, everybody magically became a bit more logical.

In that meeting, I was intent on focusing on the brutal facts. We went around the table to ensure that we got everything out. I wanted everyone's thoughts, ideas, concerns, comments—everything, every detail. When the conversation went down a rabbit hole, or threatened to become laden with opinions, I'd bring it back to the facts and figures. From the project-management standpoint, that's mostly where the concerns were: How much roofing membrane was it going to take? What was the roofing going to cost? How long was the installation going to take? How did the roofing and solar schedules overlap? As we laid the facts and figures out on the table, some folks who weren't as comfortable about things before we got together became more comfortable with talking about the reasons for their discomfort. And I think they felt the team understood where they were coming from.

Really, it was such an intense discussion that, at one point, Tom, who can be a bit of a jokester, said, "Come on, folks. Are we AmeriCANs, or AmeriCAN'Ts?"

At the end of the meeting, I went around the table one last time and asked for input on whether we should move forward with the project. I let everyone know it was okay to say no, because I would rather know about their disapproval up front than have people come back six months later and tell me that a problem we were facing was something that had worried them earlier.

In the end, everyone was onboard. All of them voted to move forward with the project.

After the meeting, the teams really began working well together. Everybody understood the initiative going forward, and communication was much better between the teams. This became crucial in overcoming some of the hurdles we faced on the project.

The first challenge that came up at IKEA Perryville was that the utility company there, Delmarva Power, imposed a maximum of two megawatts of solar energy per meter. The solution was to install multiple meters, which was possible because of the building's size, but which required two interconnect approvals from Delmarva.

Around the same time, Uni-Solar was starting to have financial trouble. Chinese manufacturers were beginning to mass-produce solar products, and they were competing heavily on price with the US manufacturers. Prices started to drop precipitously in solar energy, mostly because of the unbelievable undercutting of prices by Asian manufacturers. Since Uni-Solar was a niche product, made of amorphous silicon, it reached a point where it couldn't compete with traditional polycrystalline and monocrystalline solar panels. Uni-Solar couldn't change its manufacturing process fast enough to lower the cost of making its panels.

It was a disheartening time for us at Inovateus because we had essentially started our business model by being a Uni-Solar distributor and by learning from Stan Ovshinsky, the founder of Uni-Solar and Energy Conversion Devices. We had built a great relationship with Uni-Solar, and we were one of their go-to companies and partners.

So we had some tough decisions to make. We had just won a big contract with IKEA, and we knew that the Uni-solar product was ideal for the IKEA roof, but we weren't sure if Uni-Solar was going to be able to make or deliver the product for the job.

Eventually, we devised a plan whereby IKEA would buy the solar panels directly from Uni-Solar and warehouse them until we had the roofing work completed. This involved having our logistics team work directly with IKEA, and our purchasing team help facilitate a sale between the manufacturer and the customer. They were measures we hadn't implemented before. It took a lot of teamwork with Uni-Solar, IKEA, and others to make all this happen. As part of the deal, we agreed to work with IKEA and bring in professionals to help with the panels should anything go awry after Uni-Solar folded, because the technology wasn't going away. And in the worst case, the manner in which we installed the panels would give us options for removing or replacing them down the road.

With only one interconnect approval at the IKEA Perryville site, we moved ahead with both it and the Westampton project. Even though they were big undertakings, the projects were on schedule.

Ultimately, however, because of additional Delmarva requirements regarding meters, Perryville turned into a two-phase project.

Phase one had been completed on the first meter at Perryville, and we were pushing ahead with phase two when Uni-Solar went out of business. The company could no longer compete with the flood of products from the Asian market.

We reassessed the roof and found that the section of the roof we still needed to complete actually could meet the pounds-per-square-foot installation of traditional polycrystalline solar panels if we were to use a different racking product. We pitched the idea to IKEA, and it was approved. It took another six months to get final approval from Delmarva on the second meter, which allowed us to finally move forward and complete the project.

Our engineering team members, who had been leery about the IKEA projects in the first place, helped us overcome some major hurdles with Delmarva and persuaded the utility company to move forward with our installation. At the time, the standard inverter was rated to handle six hundred volts, but after some conversations with the utility company, we ended up using one-thousand-volt inverters, which was, then, something of a first.

So, at the end, we were able to get more power for IKEA. We did some things that hadn't been done before. We worked with one of our partners and key manufacturers as it went bankrupt and eventually shut down. And we developed new teamwork skills that really enforced our core value of *esprit de corps*.

We've since continued to work with IKEA on other projects, which is somewhat unique in the solar industry where repeat business is not all that common. I attribute that to our team spirit.

Teamwork is something we make a concerted effort to improve all the time. One of the ways we do it is through an initiative called iTEAM, which guides how we present ourselves to our customers, our vendors, our partners—everyone gets our team treatment.

We've tried to diagnose poor teamwork by looking at—among other things—what hinders our ability to effectively work as a team. And finger-pointing is definitely one of the top hindrances. As a leader, I try to guide the team to avoid finger-pointing. Instead, I try to promote a team spirit that welcomes all ideas. I like to say we rush to discovery versus rushing to judgment, and in the discovery phase, we find new things to help push our company forward.

CHAPTER 9
MOVING AHEAD

In 2012, our company was experiencing the same growth that the solar industry had been experiencing for at least seven years: we were both doubling on an annual basis. While it was a great thing, it was a challenge to manage the fast-paced growth.

Since then, the challenge has been to build a team that can implement the initiatives and the strategy that we've put in place. So we've been focused more on delivering projects to companies such as IKEA on budget and on schedule. Even more so, we provide efficient, safe products that clients can understand how to own and operate.

Near the end of 2013, the Duke Realty projects in Indianapolis came to life because Indianapolis Power & Light had instituted a solar program. At the same time, many other Midwestern utilities launched their first solar-energy programs, and interest in solar started to really spread. It continued to grow into early 2015, when we were awarded a project with the utility company in our own service area, Indiana Michigan Power.

Beyond utilities, we also began to see Midwestern opportunities for other commercial-rooftop installations, such as for IKEA in St. Louis,

Missouri, and in Canton, Michigan. In short, 2013 began an era in which we saw the Midwestern solar market start to mature. But more specifically, it started maturing in our favor. Since we had been pitching Midwestern companies and individuals to go solar, starting back in 2007, we were the first company to pop up on their radar when they actually decided to launch their initiatives. The five years we had invested in working and establishing relationships began opening doors, creating huge opportunities for us in terms of new customers and new business.

IKEA in St. Louis, Missouri

At the same time, we were working to improve our internal operations. In addition to looking for ways to improve delivery to our customers by growing our team and construction project management and engineering capabilities, we were also trying to grow the leadership team. To do that, I began to reach out to other business experts and advisers, along with other leadership groups.

Among those with whom I really hit it off was William Ditzler, who became a consultant after retiring as CEO of JFNew, an environmental company based in Indiana, which was known for its work in remediation, landscaping, and wetlands delineation. I met Will through Executive Forums. Will was someone who had become president of a company while in his midthirties, as I had. He had grown that company from a few to several hundred employees and merged it with a larger company. Since we had similar interests, I asked Will to be a facilitator for some of our leadership meetings.

Will introduced me to Gazelles International. He had just gone through certification to become a Gazelles coach. Gazelles International is led by Verne Harnish, whose credentials include serving as the contributing editor for *Fortune* magazine and authoring the book *Mastering the Rockefeller Habits*. Will suggested I read the book, which recommends using key metrics, meeting rhythms, and cascading information throughout a company. The principles in the book are based on those pioneered by John D. Rockefeller of Standard Oil, and they're put in place to make a company more efficient in its communication. From a leadership perspective, it's based on weekly meetings that cover key metrics and top priorities for the leadership team, which actually cuts down the number of meetings held in an organization.

In implementing the Rockefeller habits at Inovateus, we began holding half-hour meetings to facilitate communications, and we hold a considerably longer monthly meeting that provides a financial snapshot of the previous month along with key initiatives. In that monthly meeting, we also review larger projects, which we call rocks, and sometimes, we talk about other items such as talent, hiring, firing, top-grading, and so on—essentially a review of the team. Each quarter, we have an all-day, off-site meeting for the leadership team, which basically takes a full dive into the financials and key metrics from the previous quarter. In that quarterly meeting, we look at our estimated annual profit (EAP) report, which is based on a percentage completion of projects and where we are to date on projects. And we look at our key initiatives and our project pipeline, which includes reviewing presentations of our rocks within the team. And every year, we hold a day long planning meeting off-site, in which we come up with our one-page strategic plan and key metrics for the following year. With each of these meetings, we share the most vital takeaways with the rest of the organization.

In the fall of 2015, I had the opportunity to spend some time with Verne Harnish at the Fortune Growth Summit, which focuses on growing companies. I thanked him for sharing the Rockefeller habits information, and I shared with him the story of Inovateus, including some of the challenges and struggles that my father and the company faced. Verne told me that one of the reasons he started Gazelles International was to help entrepreneurs have access to the tools and resources they need to be successful. His father was a businessman who had experienced the extreme highs and lows of entrepreneurship, and Verne wanted to help others avoid that rollercoaster. After I shared some of our stories and my insights, I was invited to speak to the audience of more than one thousand attendees at the October 2015 event.

Other ideas we've implemented on the leadership level at Inovateus include a Friday think-tank session. Every Friday at 8:30 a.m. we get the company together for a breakfast session to talk about different topics. There are only two rules for these sessions: (1) we don't talk about what we are doing today, and (2) no idea is a bad idea. Sometimes these sessions include guest speakers, who have ranged from local community leaders, area businesspeople, and US senators to Andrew Berlin, a partner of the Chicago Cubs and owner of the South Bend Cubs; Pete Buttigieg, the mayor of South Bend and a friend of mine; Mike Brey, the head coach of the Notre Dame

Fighting Irish basketball team; Chris Stevens, cofounder of Keurig Premium Coffee Systems; and many, many others, and we got some great teamwork concepts and insights from all of them. In fact, we've been able to come up with many creative ideas in the think-tank sessions that we end up implementing within the company.

The idea for the think-tank sessions was modeled after a practice I read about in Sarah Caldicott Miller's book *Midnight Lunch: The 4 Phases of Team Collaboration Success from Thomas Edison's Lab*. Miller was Thomas Edison's great-grandniece. The concept is based on the practice that after the workday ended at Edison's company (which eventually became GE), the inventor would buy food and drink for anyone who volunteered to return to the office in the evening for a brainstorming session. Those sessions created camaraderie among the team and led to ideas that became breakthroughs at the company.

I invited Sarah to speak to a small group of Inovateus leaders, and she shared more insights on the midnight-lunch concept with us. Soon after, we started our think-tank sessions, and today we have a following of people from outside Inovateus who regularly attend those sessions. These "fans" include professors from area universities and leaders from a local utility company. It reminds me a little of my days playing in a band, when we created a fan base for our music. People in the community talk about our sessions, and others in the solar industry have heard about them. Visitors to our offices often attend the Friday sessions.

One of the most recent changes we've made is to formalize our organizational structure. We're very different from many companies; we don't follow the norm in terms of our structure and our operation. I liken us to the Grateful Dead of the music world. I'm a big fan of the band's music, but I also love the band's business innovations. They didn't follow the rules and the guidelines set forth in the music industry; they were true to their music first, and they let their processes and procedures build on that. It's the same with Willie Nelson, a musician who was considered an outlaw because he didn't fit the mold for a country musician. His music was part blues, part jazz, and so on. He was true to the music first, and upon that he built his processes and procedures.

At Inovateus, I think we have philosophies similar to the Grateful Dead and Willie Nelson. First, we've been true to building a brilliant tomorrow through clean energy, sustainable business practices, and sharing our core values of PEACE with the world. And since we focused on those first, our processes and procedures—the way we do things—have come second.

So we've created our own corporate structure, and in the latter part of 2015, we put together an organizational chart. But our chart doesn't look like the conventional pyramid-style organizational chart. Instead, it's based on Simon Sinek's concept of the golden circle, and it starts with the leaders and board in the center, which is overlapped by circles representing our projects, products, iTEAM, and ownership. Those circles are then broken down by core values and other structural pieces of the organization. The middle circle represents the *why*, as our core, and the outer circles represent the *hows*. It's more of a flat organizational chart that shows how team members interact with everybody, not just how they report to their manager.

Take, for instance, John Jackson. While he's focused on operations, engineering, and construction, the circles also demonstrate how he interacts with projects and finance and other areas of the business. The circles demonstrate how someone with ownership over one particular area may also overlap other areas and be connected with the rest of the team.

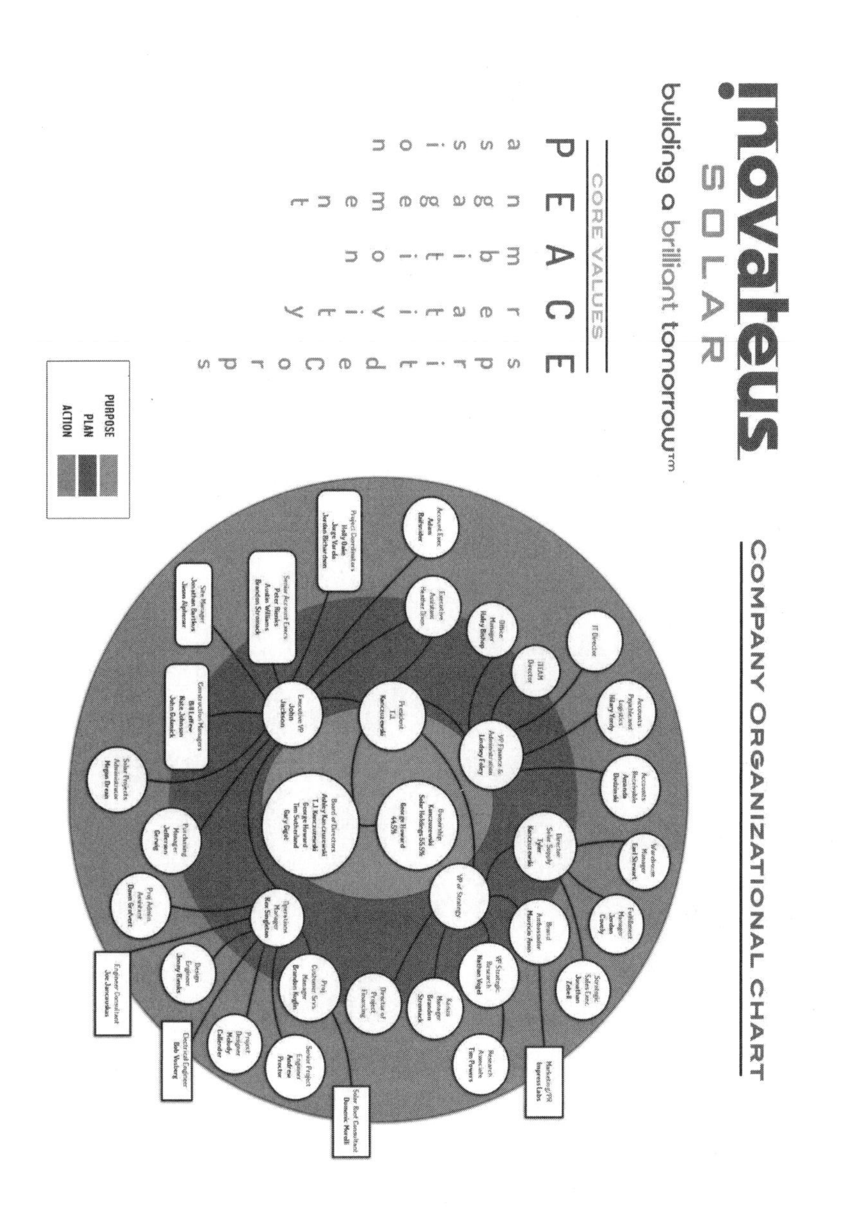

Similarly, our ten-year business plan is also somewhat unconventional in that it looks like an actual road map. It shows our geographic concentration and where we're going in the next ten years all on one page. Then we have a one-page strategic plan (see page 113) that consists of our goals for the next three to ten years. I'd like to thank Jack Daly, the well-known sales guru and a wonderful business mind and friend, for inspiring this road map (it's something he did himself). Jack was also a major inspiration for this book and I thank him dearly.

The fact that we don't fit the mold, that we're breaking with norms, is reflective of our space in the energy revolution. Our focus is solar, but we're not just revolutionizing the way that energy is delivered to customers; we're also revolutionizing the way that business is done. We hope some of what we're doing may even serve as examples for other businesses and industries.

Tyler Kanczuzewski, Nick Kanczuzewski, our good friend Jack Daly, Ashley Brickley, TJ Kanczuzewski

Moving forward, the sky is the limit. Actually, I recently learned from Buzz Aldrin, a man who set foot on the moon nearly fifty years ago, that *the sky is not the limit.* He shared with me that there are no limits to accomplishing your goals and dreams and that the sky was not a physical limit for him.

PEACE MAP, illustrated by Heather Dixon

CHAPTER 10

THE SOLAR (AND OUR) REVOLUTION

The years 2015 and 2016 have been our best to date in terms of revenue, profit, the growth of our team, the development of our leadership, and the contracts that we have been able to sign.

In November 2015, we signed our largest contract to date with DTE Energy. As I write these words, the project is roughly halfway finished, with more than 11,300 posts installed, supporting more than 3,300 racking systems equipped with nearly 68,000 solar panels. We've also begun work on the second piece of the project, a 165-acre site a few miles away that will house a twenty-four-megawatt solar-power plant using some 88,000 solar panels to harness the power of the sun. When completed, the project, totaling sixty megawatts, will be one of the largest in the United States.

The project has helped us leapfrog into the top fifteen of the nation's solar developers, and our goal is to be in the top five by the end of 2022.

In chapter 9, I detailed how Inovateus implemented the Rockefeller habits outlined by Verne Harnish to become a business poised for major growth. I mentioned our one-page strategic plan, which we've taken to the next level for our leadership and sales teams in the form of what we're calling our playbook.

As I mentioned in chapter 8, many sports teams are run better than some of the largest companies in the world because industry tends not to plan and practice and strategize in quite the way that many teams do. Often, people start work at a new company with a job description but without any real understanding of the plays the company is going to make in general, the plays of their teammates, or how everything works together.

Our playbook's purpose is similar to that of playbooks used by sports teams and, in fact, has a number of sports references within its pages. Based on the one-page strategic plan that we put together for our leadership team, I've taken that one-page plan idea and come up with some individual playbooks that include priorities for each leader's focus area. Leaders can then pass these priorities along to their respective teams.

For instance, I've created playbooks for John Jackson, executive vice president; Tom Brown, vice president of strategy; Lindsey Bauer, vice president of finance and administration; and Rex Singleton, our operations manager. Each of the books is self-titled (e.g., the *John Jackson Playbook*, the *Tom Brown Playbook*), and each aligns the leader's team goals, objectives, and priorities with our company's goals and missions. We've now worked with all team members to make sure they have a copy of the company's playbook so that everybody is literally on the same page.

John Jackson, for example, focuses on the company's one-page strategic plan, along with the goals for our operations team. Rex

Singleton is more focused on engineering and project management, so his playbook has his goals for those areas of the business.

We've also developed a playbook for our sales-and-development team, which lays out our target markets and our sales-and-development focus. Every new member of the sales-and-development team gets a copy of the playbook, which resembles an actual National Football League playbook and has everything explained in clear detail so that every member of the team knows our goals, objectives, and mission.

The playbooks all have specific goals and metrics that are supposed to be hit, and performance reviews are based on hitting those metrics. Metrics range from megawatts completed for the year to projects being on time and on budget.

The idea is that, at the end of a project, we have a happy customer who wants to promote our company to others. So the metrics we include in the playbooks are designed to meet that goal.

We also realigned our team in 2015. We had to grow the operations team significantly to deliver many of the projects that we had signed the previous year, and as a result, no one has the same job description that they did in the past. Everyone has taken on additional responsibilities, and we've been working to fine-tune our roles. Part of that reorganization was to focus our sales-and-development team on specific categories—utility, commercial, international, energy storage—in specific geographic areas.

Our primary target market is within a five-hundred-mile radius of South Bend, Indiana. That's our backyard, and we want to maximize our opportunities and make sure that we own a significant share of this market. We've worked hard to build a reputation in this marketplace, and it's really an area where we can deliver exceptionally high value because of our proximity and because our Midwestern values are well aligned with other companies in the region.

We've also had to ramp up our finance and administration team. As we've landed more projects, we've needed more financially focused individuals who can help with managing our books: accounts receivable, accounts payable, project estimation, and purchasing.

Along with our growth in 2015, we continued to reorganize our structure into pods to better address each major project. The pod idea was inspired by General Stanley McChrystal who put together a communication system that allows decisions to be made quickly. In his book *Team of Teams* McChrystal explains that gaps in our military system left our nation unprepared for attacks by hidden and isolated terrorist groups.

Recognizing our own gaps at Inovateus, we created a pod structure composed of four core people: a senior account executive, a construction manager, a project coordinator, and a project engineer, although anyone in the organization can be part of a pod and attend meetings involving the project. As was its intent, the pod structure has worked very well on major initiatives—for instance, the DTE Energy project I mentioned earlier.

We also invested in the education of our team to strengthen our organization for the days ahead, and we're looking forward to seeing many of our current team members become leaders at our company. We have a program that pays 100 percent of tuition for any of our employees to receive education that can help them in their career. At the writing of this book, a member of our sales-and-development team is enrolled in the Notre Dame Executive MBA program, which is one of the top executive masters of business administration programs in the country. Two of our engineering employees are getting their engineering degrees, and one engineer is in the process of getting his professional engineering (PE) seal. In addition, John Jackson attended a weeklong executive-education program at Harvard in 2015, and Tom Brown is working on a masters in business dynamics through Gazelles International.

By the end of 2015, I had started the conceptual formation of our company's first external board of directors, and we are going to conduct our first meeting at the end of 2016. I have put together a board of directors because our current ownership group doesn't have the experience to guide our leadership team through certain initiatives, such as the very large construction projects Inovateus is taking on. With my father's passing, my five siblings and I became 55 percent owners of Inovateus Solar, with 45 percent owned by George Howard. I felt the new ownership team needed the support of a board of directors composed of people who have experience in running successful companies or in putting together creative business strategies or who have expert legal and financial backgrounds.

After an extensive search, I located five individuals with diverse skill sets and experience to guide and review the performance of our executive team, which includes me. I expect this to be a big help for us in the years to come as our revenue doubles and triples.

While we worked on our people and processes throughout 2015, the solar industry underwent many changes as well. Battery-storage technology has become far more cost effective. Now, when a solar-energy system is married with efficient battery storage, it makes for far greater opportunities for independence from the electricity grid.

Late in 2015, a phenomenal occurrence took place in Washington, DC. A 30 percent tax credit for installing solar on residential and commercial properties, in place since 2006, was scheduled to expire at the end of 2016. Upon its expiration, it would be replaced by a 10 percent tax credit on January 1, 2017. The expiration date had many large financial institutions starting to back away from solar as an investment at the end of 2015. Many large projects take over a year to get started, but most of our contracts for 2016 were finalized. But the future beyond that was a little uncertain.

Knowing that the deadline was coming, we'd been working on plans to continue growth into 2017. That included working very hard with our local representatives in Congress. Since our headquarters is on Stateline Road, our offices are actually in Indiana, but just across the street is Michigan (half of our employees live in Indiana, and half live in Michigan). So in addition to meeting with US Senator Joe Donnelly and US Representative Jackie Walorski from our own district in Indiana, we've met with US Representative Fred Upton, the representative for Michigan's Sixth Congressional District, who is also the chairman of the House Committee on Energy and Commerce.

I'd like to think our voices were heard because on the day I sat down to write this chapter, very positive progress on restoring the 30 percent tax credit was announced. Incentives were added to the 2016 Omnibus Appropriations bill, a $1.15 trillion federal-spending bill that had strong bipartisan support. The bill passed on a vote of 316–113, extending the investment tax credit (ITC) of 30 percent through 2019. The ITC applies to residential and commercial installations, including utility companies. After 2019, the credit will begin a decline, dropping to 26 percent in 2020, 22 percent in 2021, and 10 percent in 2022.[4] Any project that commences construction before the end of 2021 but places that project in service before the end of 2023 will be grandfathered into the system.

4 "Solar Industry Expected to Add over 200,000 New Jobs by 2020," Solar Energy Industries Association, accessed January 7, 2016, http://www.seia.org/research-resources/impacts-solar-investment-tax-credit-extension.

The extension of the tax credit was great news for the solar industry and the solar revolution. Overnight, stocks of major players in the industry shot up by record amounts. It was also great news for subsequent five-year sales outlooks for manufacturers in the United States. To reiterate and expand on points made earlier by Rhone Resch, the former president of the Solar Energy Industries Association (SEIA): "By 2020, the industry will deploy more than 20 gigawatts (GW) of solar electric capacity annually and employ more than 420,000 workers. The additional solar generation will more than offset carbon emissions from the lift of the oil export ban on an annual basis by 2019."[5]

With the expiration of the ITC originally set for January 31, 2016, many individuals and companies were trying to ensure that their installations were in place. Consequently, there was a big push to have a lot of work done by the end of 2016. Because of that push, a lot of companies like ours had plenty of work on their plate, enough to make for a breakthrough year—and potentially enough to triple or quadruple the installed capacity across the US. That equates to advancements in technology and the way solar energy is delivered to the customers. We're going to see products become more efficient, from more efficient solar panels and inverters to advancements in ground-mount tracking systems that follow the sun throughout the day to maximize the harvest of sunlight.

Not only were 2015 and 2016 productive years, they were also very exciting.

In February 2016, I was invited to go to the White House to address the president's cabinet members on the Clean Power Plan being discussed in Congress. They wanted to get a Midwestern view of how clean energy was being implemented, whether it was

5 Ibid.

working, and how the Clean Power Plan could help with any initiatives. Obviously, it was an honor to go to the White House and represent Inovateus Solar, and it was an honor to meet and speak to President Barack Obama.

Beforehand, I wanted to get a full understanding of the Clean Power Plan. So I dug into the plan of over four hundred pages and read up on its history. In doing so, I discovered that while the plan supported the transition to cleaner forms of energy such as solar, wind, hydropower, geothermal, battery storage, and so on, it didn't really embrace job creation in the US. At the time, the majority of solar-technology makers were located overseas.

I also wanted to get the viewpoint of our congressional representatives regarding the Clean Power Plan. So I visited the offices of Senator Donnelly and Representatives Walorski and Upton. But I found that each of them was opposed to the plan because they thought it would increase the cost of energy for Indiana businesses and homeowners, which would also make Indiana and Michigan less competitive when it comes to creating jobs. So I explained to them that while the Clean Energy Plan requires states to meet clean-energy thresholds, each state is allowed to come up with its own way to govern the initiative. In

TJ Kanczuzewski at the White House

essence, each state is allowed to put together the ground rules that work best for it.

Armed with a deeper understanding of the plan, I went to Washington, DC. While walking around the National Mall, I got the sense that the US government is more than a political entity; it is, in essence, one of the largest companies in the world. So in the meeting at the White House—which, by the way, meant passing through a very impressive, multilayered security system—I did my best to shed light on the fact that our state's congressional representatives didn't understand the Clean Energy Plan's potential for creating jobs. I explained that there was a disconnect between the plan's verbiage and what many understood as its intent.

In some small way, I believe I helped to bridge a gap in understanding the plan. And I've since tried to arrange a meeting to explain how the infrastructure of Indiana and Michigan are perfectly suited to the manufacture of renewable-energy products.

After I returned to Indiana from my meeting at the White House, I was selected as Outstanding Young Business Leader by the St. Joseph County Chamber of Commerce and 1st Source Bank at the 2016 Salute to Business luncheon. The award goes to community business leaders who have reached a heightened level of achievement, show promise for ongoing success, engage other young professionals, and have had an impact on their community. I was truly honored to be chosen

TJ Kanczuzewski accepting Outstanding Young Business Leader award

for such an important distinction, as it was a reflection of the outstanding team members and people I get to work with every day.

Between January and March, we at Inovateus were pleased to welcome professional IndyCar driver Stefan Wilson and his team to our offices. Stefan is passionate about renewable and clean energy and told us about a campaign he was kicking off called #ThinkSolar. We had met with him earlier, in December 2015, and were intrigued then by the idea that someone who burns fossil fuels for a living could be passionate about renewables and even wanted to see the advancement of electric vehicles in car racing.

Stefan talked to us about partnering with him to promote solar energy at the Indianapolis Motor Speedway. A seasoned Indy Lights driver, Stefan was hoping to qualify for the hundredth anniversary of the Indianapolis 500 race to further his career and also to pay tribute to Justin Wilson, his older brother and a successful and well-loved IndyCar driver, who passed after a tragic incident during a race in 2015.

I admired Stefan not only for his passion about solar but also for his courage and ambition to continue the legacy of his brother and mentor—something similar to what I was doing at Inovateus. As I've mentioned in this book, I've tried to continue my father's legacy of promoting renewable energy and creating, in South Bend, Indiana, a great business that supports many people.

A few weeks later, I met up with Stefan and his manager, Anders Krohn, a former driver who runs CoForce International. Anders manages other drivers and is a color commentator for IndyCar events. He has many social/business connections, and some consider him to be a god in the racing world. We met at the South by Southwest conference in Austin, Texas. There, it was interesting to learn more about Stefan not only as a genuinely humble person but also as a confident risk taker. After all, he drives a car around a track at over two hundred miles per hour. When I got to know him better, Inovateus inked a deal to partner with him in early April, and shortly after that, he was accepted as a driver in the hundredth celebration of the Indy 500. Several members of the Inovateus team went to watch him on a practice day prior to the big event, where he launched his #ThinkSolar campaign. And we were definitely at the Indy 500 race on Memorial weekend, along with 370,000 attendees, watching Stefan's car, bearing the Inovateus logo, round the track. It was amazing! My brother Nick, who is a racing enthusiast, was able to join the Inovateus team and enjoy this exciting day we will never forget.

Although Stefan started off the race in a very strong position, he ended up, unfortunately, finishing near the back of the pack because of technical difficulties. But we were proud of him all the same, and we see a great future for him and for our partnership, which continues online.

In mid-2016, after overcoming some tax-abatement issues, we finally began work on another major project, the thirteen-megawatt, solarized, parking canopy at Michigan State University (MSU) that I mentioned in chapter 7. Equipped with more than forty thousand

Stefan Wilson with Inovateus team members

solar panels, the custom-built arrays are in five locations on the south end of the now coal-free campus and, together, comprise one of the largest photovoltaic arrays at any US university.

In July, Inovateus Solar once again participated in the Intersolar North America conference, where we joined with other solar professionals from around the world to talk technology, negotiate deals, and discuss our industry. While in the San Francisco Bay Area for the conference, I managed to set up a lengthy one-on-one meeting with Phil, a founding member and the bass player for the Grateful Dead and several other groups, including Phil Lesh and Friends, the Other Ones, and Further.

As I mentioned in chapter 9, I'm a big fan of the Grateful Dead and have always been intrigued by the band's business practices. In talking with Phil, I hoped to gain some insights into the Dead's business philosophy and see if there was something I could apply to Inovateus. He told me that, in business or life, you must be a good listener. "You must listen to the universe," he said, "because the answers in life are already out there if you're just open to hearing them. Sometimes," he continued, "the universe blurts out what you need to know. But at other times, you must listen carefully because the universe also speaks in whispers."

Listening, Phil said, is what led him to follow his calling as a musician and to meet his future band mates, including Jerry Garcia. And being a listener became a sort of critical, unspoken, yet understood virtue that the Grateful Dead shared onstage and in their business practices. They listened to their fans, the music industry, promoters, and the 1960s counterculture, and in the process, they created an almost tribal community. Phil explained that listening was also something that held the band together during periods of disagreement because when the band took the stage and performed as one, listening to each other, the problems would go away for a while. He described listening as a sort of healing process. "I'm not sure how you could apply something like that to your business," he said, "but it's something to ponder."

That led me to talk with Phil about how Inovateus has been a company of "strategic improvisation" when pursuing ideas, new directions. Phil's advice, as a master of improvisation, was to nudge ideas rather than strong-arm them. That's how the Grateful Dead improvised, he explained: It was the nature of the band to join in when a song was being nudged in a new direction, to basically accept the invitation to play along. But a new direction for a song, in midplay, could never be forced. The same concept applied to ideas for concerts, merchandise, and working with vendors or the folks whom they allowed to tape their shows, known as tapers.

I talked to Phil about the band's business practice of state-of-the-art sound systems to ensure quality audio at its concerts, pioneering digital effects, and its early and ongoing use of online technologies to build a cult-like fan base. Much of the band's cutting-edge ideas were driven by the desire to control its own destiny, including creating its own record label and making ground-breaking use of its own sound systems while on the road, instead of following the standard practice in the late 1960s of using equipment supplied by the performance venues.

But being cutting edge was secondary to having a passion for music, Phil said. Above all, the band was driven by a passion for music; that desire was at the forefront of its efforts. For instance, the reason for innovations such as the band's Wall of Sound was to give fans the best experience possible.

When Phil shared his story with me, I saw similarities with our own story at Inovateus. Everything we've done is out of our passion to keep solar at the forefront and to control our own destiny.

Later that evening, our band from Inovateus once again played in the Solar Battle of the Bands. This time, we took first place in the competition. And the following evening, we had a company party at Terrapin Crossroads, a club owned by Phil Lesh, where we're now installing a state-of-the-art solar-energy system. It's been a trip to work with Phil and the team at Terrapin to provide solar power for the endless music performed at the club.

Phil Lesh

Battle of the Bands

To quote Grateful Dead lyricist Robert Hunter in their classic song "Truckin": "Lately it occurs to me, what a long strange trip it's been." Needless to say, it's been an amazing and challenging "long strange trip" for us at Inovateus Solar and for the solar industry as a whole. As you undoubtedly have discovered, solar is my passion. I can, and do, talk about it from sunup to sundown and beyond. But this story has to end somewhere. So I've reached a conclusion to our written history. But our story will continue alongside solar's vibrant future, which will lead us to new roads when the time is right.

CONCLUSION
OUR BRIGHTER TOMORROW

As Inovateus has grown on its journey to become a world-class solar company, one of the most important things I've learned and want to share with other young leaders is this: No matter who you are or what you do, be yourself. That's the one thing in the world you can do better than anyone else. When you find something that you're really passionate about, that you really believe in, you've just got to go with it.

Back in 2007, when I made the decision to work at Inovateus, it was one of the best decisions I ever made in my life. When I made that decision, I didn't know what the future held. But solar was my passion, and all I was doing was trying to be me—I was trying to be TJ.

Today, I can still say that I would make the same decision all over again.

Recently, I rediscovered our *why*. My father started Inovateus Development, the predecessor to Inovateus Solar, based on its *why*. And when my father met George Howard with whom he became a partner at Inovateus Solar, along with my grandfather Leonard, they continued that *why*. When they were developing concepts and researching renewable energy, they always had in mind the *why* of what they were doing.

I've talked about it with the Inovateus Solar team, over the years, but as we've begun to bring in very big projects with big dollar amounts and big profits, money seems to have become more of the basis of our conversations and even the basis of some of our decisions. While profits are important, they are more of the *how* and the *what* we do, not the *why*.

So I sat down to try to put that *why*—why did we get started?—into a phrase. Obviously, Inovateus and solar are my passions, so I could talk about them for hours. But I remember when Inovateus Development got started, and even through the start-up of Inovateus Solar, my father used to say, "Live the question; promote the solution."

I meditated on that for a time because even though I thought I understood what he meant back then, I wondered if there were really more to it. I wondered how it might tie into our *why*.

So I started thinking more about the solution idea in my father's words. What solution are we trying to promote through Inovateus Solar? What was my father's mission, and what hunger did he, George, others, and I have that drove us to do what we do today? What are we trying to promote? Then it came to me. I'm a pretty busy guy, so sometimes I exercise late at night after my wife and kids have gone to sleep. And late one night, after I had finished my exercise session, I was sitting there by myself, pondering what it was that we were promoting when I heard the answer in my mind. I could almost hear my father's voice: "Build a brilliant tomorrow."

Merriam-Webster defines *brilliant* as "very bright," "very impressive or successful," and "extremely intelligent." For Inovateus, *brilliant* means a brighter tomorrow. It means a cleaner, safer world with a smarter energy source—solar today and who knows what tomorrow. And, of course, we hope it means an even greater level of success for us at Inovateus.

Whatever advances in solar, clean energy, and energy storage bring, if we continue to focus on our *why*, we will be able to build that brighter tomorrow—a brilliant future for everyone we touch with solar. Come and join us on our journey.

ABOUT THE AUTHOR

Mr. Kanczuzewski is a founding member and president of Inovateus Solar LLC and has been with the company since 2007. Driven by the company's purpose to "build a brilliant tomorrow," he has been instrumental in growing a world-class team that has successfully deployed solar-energy projects and products internationally. Mr. Kanczuzewski was honored as a 2013 Forty Under 40 recipient for northern Indiana and southwest Michigan. He is very honored that Inovateus was recognized as one of the Indiana Chamber of Commerce's 2014 and 2016 Best Places to Work in Indiana. Most recently, in February 2016, Mr. Kanczuzewski was invited to the White House to participate in a briefing on clean energy and also received the St. Joseph County Chamber of Commerce's 2016 Outstanding Young Business Leader award. He is also a board member of Fernwood Botanical Garden. Prior to his career at Inovateus Solar, Mr. Kanczuzewski worked for General Growth Properties in Chicago, Illinois. He graduated from Columbia College in 2004 with a bachelor of arts degree in business management.